Springer Texts in Statistics

Springer Texts in Statistics

Alfred	Elements of Statistics for the Life and Social Sciences
Berger	An Introduction to Probability and Stochastic Processes
Blom	Probability and Statistics: Theory and Applications
Chow and Teicher	Probability Theory: Independence, Interchangeability, Martingales, Second Edition
Christensen	Plane Answers to Complex Questions: The Theory of Linear Models
Christensen	Linear Models for Multivariate, Time Series, and Spatial Data
Christensen	Log-Linear Models
Creighton	A First Course in Probability Models and Statistical Inference
du Toit, Steyn and Stumpf	Graphical Exploratory Data Analysis
Finkelstein and Levin	Statistics for Lawyers
Jobson	Applied Multivariate Data Analysis, Volume I: Regression and Experimental Design
Jobson	Applied Multivariate Data Analysis, Volume II: Categorical and Multivariate Methods
Kalbfleisch	Probability and Statistical Inference, Volume I: Probability, Second Edition
Kalbfleisch	Probability and Statistical Inference, Volume II: Statistical Inference, Second Edition
Karr	Probability

Continued at end of book

Texts in Applied Mathematics 4

Springer Science+Business Media, LLC

Texts in Applied Mathematics

1. *Sirovich:* Introduction to Applied Mathematics.
2. *Wiggins:* Introduction to Applied Nonlinear Dynamical Systems and Chaos.
3. *Hale/Koçak:* Dynamics and Bifurcations.
4. *Chorin/Marsden:* A Mathematical Introduction to Fluid Mechanics, 3rd ed.
5. *Hubbard/West:* Differential Equations: A Dynamical Systems Approach: Ordinary Differential Equations.
6. *Sontag:* Mathematical Control Theory: Deterministic Finite Dimensional Systems, 2nd ed.
7. *Perko:* Differential Equations and Dynamical Systems, 2nd ed.
8. *Seaborn:* Hypergeometric Functions and Their Applications.
9. *Pipkin:* A Course on Integral Equations.
10. *Hoppensteadt/Peskin:* Mathematics in Medicine and the Life Sciences.
11. *Braun:* Differential Equations and Their Applications, 4th ed.
12. *Stoer/Bulirsch:* Introduction to Numerical Analysis, 2nd ed.
13. *Renardy/Rogers:* A First Graduate Course in Partial Differential Equations.
14. *Banks:* Growth and Diffusion Phenomena: Mathematical Frameworks and Applications.
15. *Brenner/Scott:* The Mathematical Theory of Finite Element Methods.
16. *Van de Velde:* Concurrent Scientific Computing.
17. *Marsden/Ratiu:* Introduction to Mechanics and Symmetry, 2nd ed.
18. *Hubbard/West:* Differential Equations: A Dynamical Systems Approach: Higher-Dimensional Systems.
19. *Kaplan/Glass:* Understanding Nonlinear Dynamics.
20. *Holmes:* Introduction to Perturbation Methods.
21. *Curtain/Zwart:* An Introduction to Infinite-Dimensional Linear Systems Theory.
22. *Thomas:* Numerical Partial Differential Equations: Finite Difference Methods.
23. *Taylor:* Partial Differential Equations: Basic Theory.
24. *Merkin:* Introduction to the Theory of Stability of Motion.
25. *Naber:* Topology, Geometry, and Gauge Fields: Foundations.
26. *Polderman/Willems:* Introduction to Mathematical Systems Theory: A Behavioral Approach.
27. *Reddy:* Introductory Functional Analysis with Applications to Boundary-Value Problems and Finite Elements.
28. *Gustafson/Wilcox:* Analytical and Computational Methods of Advanced Engineering Mathematics.
29. *Tveito/Winther:* Introduction to Partial Differential Equations: A Computational Approach.
30. *Gasquet/Witomski:* Fourier Analysis and Applications: Filtering, Numerical Computation, Wavelets.
31. *Brémaud:* Markov Chains: Gibbs Fields, Monte Carlo Simulation, and Queues.
32. *Durran:* Numerical Methods for Wave Equations in Geophysical Fluid Dynamics.

(continued after index)

Alexandre J. Chorin
Jerrold E. Marsden

A Mathematical Introduction to Fluid Mechanics

Third Edition

With 87 Illustrations

Springer

Alexandre J. Chorin
Department of Mathematics
University of California
Berkeley, CA 94720
USA

Jerrold E. Marsden
Control and Dynamical Systems 107-81
Caltech
Pasadena, CA 91125
USA

Editors

J.E. Marsden
Control and Dynamical Systems 107-81
Caltech
Pasadena, CA 91125
USA

L. Sirovich
Division of Applied Mathematics
Brown University
Providence, RI 02912
USA

M. Golubitsky
Department of Mathematics
University of Houston
Houston, TX 77004
USA

Cover Illustration: A computer simulation of a shock diffraction by a pair of cylinders, by John Bell, Phillip Colella, William Crutchfield, Richard Pember, and Michael Welcome.

Mathematics Subject Classification (1980): 76-01, 76C05, 76D05, 76D10, 76N15

Library of Congress Cataloging-in-Publication Data
Chorin, Alexandre Joel.
A mathematical introduction to fluid mechanics / A.J. Chorin, J.E. Marsden.—3rd ed.
p. cm.—(Texts in applied mathematics ; 4)
Includes bibliographical references and index.

1. Fluid mechanics. I. Marsden, Jerrold E. II. Title. III. Series.
QA901.C53 1992
532 dc20 92-26645

Printed on acid-free paper.

Production managed by Christin R. Ciresi; manufacturing supervised by Vincent Scelta.
Typesetting and graphics by June Meyermann, Ithaca, NY.

9 8 7 6 5 4 (Corrected fourth printing, 2000)

ISBN 978-1-4612-6934-2 ISBN 978-1-4612-0883-9 (eBook)
DOI 10.1007/ 978-1-4612-0883-9 SPIN 10764957

Series Preface

Mathematics is playing an ever more important role in the physical and biological sciences, provoking a blurring of boundaries between scientific disciplines and a resurgence of interest in the modern as well as the classical techniques of applied mathematics. This renewal of interest, both in research and teaching, has led to the establishment of the series: *Texts in Applied Mathematics (TAM)*.

The development of new courses is a natural consequence of a high level of excitement on the research frontier as newer techniques, such as numerical and symbolic computer systems, dynamical systems, and chaos, mix with and reinforce the traditional methods of applied mathematics. Thus, the purpose of this textbook series is to meet the current and future needs of these advances and encourage the teaching of new courses.

TAM will publish textbooks suitable for use in advanced undergraduate and beginning graduate courses, and will complement the *Applied Mathematical Sciences (AMS)* series, which will focus on advanced textbooks and research level monographs.

Preface

This book is based on a one-term course in fluid mechanics originally taught in the Department of Mathematics of the University of California, Berkeley, during the spring of 1978. The goal of the course was not to provide an exhaustive account of fluid mechanics, nor to assess the engineering value of various approximation procedures. The goals were:

- to present some of the basic ideas of fluid mechanics in a mathematically attractive manner (which does not mean "fully rigorous");
- to present the physical background and motivation for some constructions that have been used in recent mathematical and numerical work on the Navier–Stokes equations and on hyperbolic systems; and
- to interest some of the students in this beautiful and difficult subject.

This third edition has incorporated a number of updates and revisions, but the spirit and scope of the original book are unaltered.

The book is divided into three chapters. The first chapter contains an elementary derivation of the equations; the concept of vorticity is introduced at an early stage. The second chapter contains a discussion of potential flow, vortex motion, and boundary layers. A construction of boundary layers using vortex sheets and random walks is presented. The third chapter contains an analysis of one-dimensional gas flow from a mildly modern point of view. Weak solutions, Riemann problems, Glimm's scheme, and combustion waves are discussed.

The style is informal and no attempt is made to hide the authors' biases and personal interests. Moreover, references are limited and are by no means exhaustive. We list below some general references that have been useful for us and some that contain fairly extensive bibliographies. References relevant to specific points are made directly in the text.

R. Abraham, J. E. Marsden, and T. S. Ratiu [1988] *Manifolds, Tensor Analysis and Applications*, Springer-Verlag: Applied Mathematical Sciences Series, Volume **75**.

G. K. Batchelor [1967] *An Introduction to Fluid Dynamics*, Cambridge Univ. Press.

G. Birkhoff [1960] *Hydrodynamics, a Study in Logic, Fact and Similitude*, Princeton Univ. Press.

A. J. Chorin [1976] *Lectures on Turbulence Theory*, Publish or Perish.

A. J. Chorin [1989] *Computational Fluid Mechanics*, Academic Press, New York.

A. J. Chorin [1994] *Vorticity and Turbulence*, Applied Mathematical Sciences, **103**, Springer-Verlag.

R. Courant and K. O. Friedrichs [1948] *Supersonic Flow and Shock Waves*, Wiley-Interscience.

P. Garabedian [1960] *Partial Differential Equations*, McGraw-Hill, reprinted by Dover.

S. Goldstein [1965] *Modern Developments in Fluid Mechanics*, Dover.

K. Gustafson and J. Sethian [1991] *Vortex Flows*, SIAM.

O. A. Ladyzhenskaya [1969] *The Mathematical Theory of Viscous Incompressible Flow*, Gordon and Breach.

L. D. Landau and E. M. Lifshitz [1968] *Fluid Mechanics*, Pergamon.

P. D. Lax [1972] *Hyperbolic Systems of Conservation Laws and the Mathematical Theory of Shock Waves*, SIAM.

A. J. Majda [1986] *Compressible Fluid Flow and Systems of Conservation Laws in Several Space Variables*, Springer-Verlag: Applied Mathematical Sciences Series **53**.

J. E. Marsden and T. J. R. Hughes [1994] *The Mathematical Foundations of Elasticity*, Prentice-Hall, 1983. Reprinted with corrections, Dover, 1994.

J. E. Marsden and T. S. Ratiu [1994] *Mechanics and Symmetry*, Texts in Applied Mathematics, **17**, Springer-Verlag.

R. E. Meyer [1971] *Introduction to Mathematical Fluid Dynamics*, Wiley, reprinted by Dover.

K. Milne–Thomson [1968] *Theoretical Hydrodynamics*, Macmillan.

C. S. Peskin [1976] *Mathematical Aspects of Heart Physiology*, New York Univ. Lecture Notes.

S. Schlichting [1960] *Boundary Layer Theory*, McGraw-Hill.

L. A. Segel [1977] *Mathematics Applied to Continuum Mechanics*, Macmillian.

J. Serrin [1959] Mathematical Principles of Classical Fluid Mechanics, *Handbuch der Physik*, **VIII/1**, Springer-Verlag.

R. Temam [1977] *Navier–Stokes Equations*, North-Holland.

We thank S. S. Lin and J. Sethian for preparing a preliminary draft of the course notes—a great help in preparing the first edition. We also thank O. Hald and P. Arminjon for a careful proofreading of the first edition and to many other readers for supplying both corrections and support, in particular V. Dannon, H. Johnston, J. Larsen, M. Olufsen, and T. Ratiu and G. Rublein. These corrections, as well as many other additions, some exercises, updates, and revisions of our own have been incorporated into the second and third editions. Special thanks to Marnie McElhiney for typesetting the second edition, to June Meyermann for typesetting the third edition, and to Greg Kubota and Wendy McKay for updating the third edition with corrections.

ALEXANDRE J. CHORIN
Berkeley, California

JERROLD E. MARSDEN
Pasadena, California

Summer, 1997

Contents

Preface vii

1 The Equations of Motion 1

1.1 Euler's Equations 1
1.2 Rotation and Vorticity 18
1.3 The Navier–Stokes Equations 31

2 Potential Flow and Slightly Viscous Flow 47

2.1 Potential Flow 47
2.2 Boundary Layers 67
2.3 Vortex Sheets 82
2.4 Remarks on Stability and Bifurcation 96

3 Gas Flow in One Dimension 103

3.1 Characteristics 103
3.2 Shocks 117
3.3 The Riemann Problem 137
3.4 Combustion Waves 145

Index 165

1

The Equations of Motion

In this chapter we develop the basic equations of fluid mechanics. These equations are derived from the conservation laws of mass, momentum, and energy. We begin with the simplest assumptions, leading to Euler's equations for a perfect fluid. These assumptions are relaxed in the third section to allow for viscous effects that arise from the molecular transport of momentum. Throughout the book we emphasize the intuitive and mathematical aspects of vorticity; this job is begun in the second section of this chapter.

1.1 Euler's Equations

Let D be a region in two- or three-dimensional space filled with a fluid. Our object is to describe the motion of such a fluid. Let $\mathbf{x} \in D$ be a point in D and consider the particle of fluid moving through $\mathbf{x}$ at time t. Relative to standard Euclidean coordinates in space, we write $\mathbf{x} = (x, y, z)$. Imagine a particle (think of a particle of dust suspended) in the fluid; this particle traverses a well-defined trajectory. Let $\mathbf{u}(\mathbf{x}, t)$ denote the velocity of the particle of fluid that is moving through $\mathbf{x}$ at time t. Thus, for each fixed time, $\mathbf{u}$ is a vector field on D, as in Figure 1.1.1. We call $\mathbf{u}$ the (***spatial***) ***velocity field of the fluid.***

For each time t, assume that the fluid has a well-defined ***mass density*** $\rho(\mathbf{x}, t)$. Thus, if W is any subregion of D, the mass of fluid in W at time t

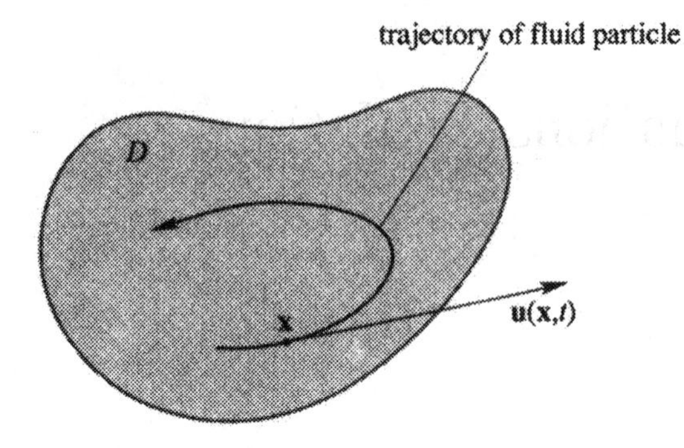

FIGURE 1.1.1. Fluid particles flowing in a region D.

is given by

$$m(W,t) = \int_W \rho(\mathbf{x},t)\,dV,$$

where dV is the volume element in the plane or in space.

In what follows we shall assume that the functions $\mathbf{u}$ and ρ (and others to be introduced later) are smooth enough so that the standard operations of calculus may be performed on them. This assumption is open to criticism and indeed we shall come back and analyze it in detail later.

The assumption that ρ exists is a ***continuum assumption***. Clearly, it does not hold if the molecular structure of matter is taken into account. For most macroscopic phenomena occurring in nature, it is believed that this assumption is extremely accurate.

Our derivation of the equations is based on three basic principles:

i *mass is neither created nor destroyed*;

ii *the rate of change of momentum of a portion of the fluid equals the force applied to it* (***Newton's second law***);

iii *energy is neither created nor destroyed.*

Let us treat these three principles in turn.

i Conservation of Mass

Let W be a fixed subregion of D (W does *not* change with time). The rate of change of mass in W is

$$\frac{d}{dt}m(W,t) = \frac{d}{dt}\int_W \rho(\mathbf{x},t)\,dV = \int_W \frac{\partial \rho}{\partial t}(\mathbf{x},t)\,dV.$$

Let ∂W denote the boundary of W, assumed to be smooth; let $\mathbf{n}$ denote the unit outward normal defined at points of ∂W; and let dA denote the area element on ∂W. The volume flow rate across ∂W per unit area is $\mathbf{u} \cdot n$ and the mass flow rate per unit area is $\rho\mathbf{u} \cdot n$ (see Figure 1.1.2).

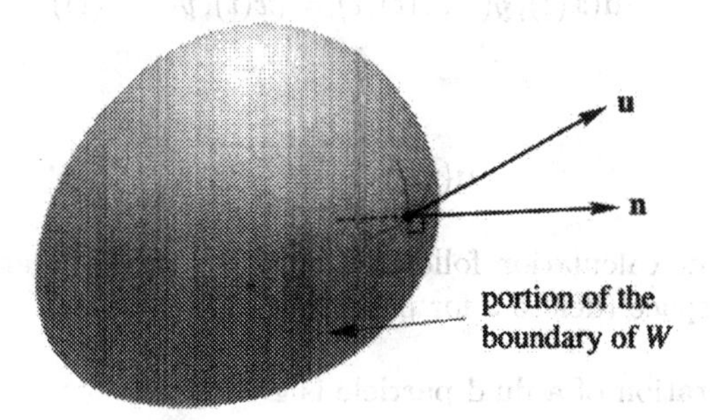

FIGURE 1.1.2. The mass crossing the boundary ∂W per unit time equals the surface integral of $\rho\mathbf{u} \cdot \mathbf{n}$ over ∂W.

The principle of conservation of mass can be more precisely stated as follows: The rate of increase of mass in W equals the rate at which mass is crossing ∂W in the *inward* direction; *i.e.*,

$$\frac{d}{dt} \int_W \rho \, dV = -\int_{\partial W} \rho\mathbf{u} \cdot \mathbf{n} \, dA.$$

This is the ***integral form of the law of conservation of mass.*** By the divergence theorem, this statement is equivalent to

$$\int_W \left[\frac{\partial \rho}{\partial t} + \operatorname{div}(\rho\mathbf{u}) \right] dV = 0.$$

Because this is to hold for all W, it is equivalent to

$$\frac{\partial \rho}{\partial t} + \operatorname{div}(\rho\mathbf{u}) = 0.$$

The last equation is the ***differential form of the law of conservation of mass***, also known as the ***continuity equation.***

If ρ and $\mathbf{u}$ are not smooth enough to justify the steps that lead to the differential form of the law of conservation of mass, then the integral form is the one to use.

ii Balance of Momentum

Let $\mathbf{x}(t) = (x(t), y(t), z(t))$ be the path followed by a fluid particle, so that the velocity field is given by

$$\mathbf{u}(x(t), y(t), z(t), t) = (\dot{x}(t), \dot{y}(t), \dot{z}(t)),$$

that is,

$$\mathbf{u}(\mathbf{x}(t), t) = \frac{d\mathbf{x}}{dt}(t).$$

This and the calculation following explicitly use standard Euclidean coordinates in space (delete z for plane flow).[1]

The acceleration of a fluid particle is given by

$$\mathbf{a}(t) = \frac{d^2}{dt^2}\mathbf{x}(t) = \frac{d}{dt}\mathbf{u}(x(t), y(t), z(t), t).$$

By the chain rule, this becomes

$$\mathbf{a}(t) = \frac{\partial \mathbf{u}}{\partial x}\dot{x} + \frac{\partial \mathbf{u}}{\partial y}\dot{y} + \frac{\partial \mathbf{u}}{\partial z}\dot{z} + \frac{\partial \mathbf{u}}{\partial t}.$$

Using the notation

$$\mathbf{u}_x = \frac{\partial \mathbf{u}}{\partial x}, \quad \mathbf{u}_t = \frac{\partial \mathbf{u}}{\partial t}, \quad \text{etc.},$$

and

$$\mathbf{u}(x, y, z, t) = (u(x, y, z, t), v(x, y, z, t), w(x, y, z, t)),$$

we obtain

$$\mathbf{a}(t) = u\mathbf{u}_x + v\mathbf{u}_y + w\mathbf{u}_z + \mathbf{u}_t,$$

which we also write as

$$\mathbf{a}(t) = \partial_t \mathbf{u} + \mathbf{u} \cdot \nabla \mathbf{u},$$

---

[1] Care must be used if other coordinate systems (such as spherical or cylindrical) are employed. Other coordinate systems can be handled in two ways: first, one can proceed more intrinsically by developing intrinsic (i.e., coordinate free) formulas that are valid in any coordinate system, or, second, one can do all the derivations in Euclidean coordinates and transform final results to other coordinate systems at the end by using the chain rule. The second approach is clearly faster, although intellectually less satisfying. See Abraham, Marsden and Ratiu [1988] (listed in the front matter) for information on the former approach. For reasons of economy we shall do most of our calculations in standard Euclidean coordinates.

where

$$\partial_t \mathbf{u} = \frac{\partial \mathbf{u}}{\partial t} \quad \text{and} \quad \mathbf{u} \cdot \nabla = u\frac{\partial}{\partial x} + v\frac{\partial}{\partial y} + w\frac{\partial}{\partial z}.$$

We call

$$\frac{D}{Dt} = \partial_t + \mathbf{u} \cdot \nabla$$

the ***material derivative***; it takes into account the fact that the fluid is moving and that the positions of fluid particles change with time. Indeed, if $f(x, y, z, t)$ is any function of position and time (scalar or vector), then by the chain rule,

$$\boxed{\frac{d}{dt} f(x(t), y(t), z(t), t) = \partial_t f + \mathbf{u} \cdot \nabla f = \frac{Df}{Dt}(x(t), y(t), z(t), t).}$$

For any continuum, forces acting on a piece of material are of two types. First, there are forces of *stress*, whereby the piece of material is acted on by forces across its surface by the rest of the continuum. Second, there are external, or body, forces such as gravity or a magnetic field, which exert a force per unit volume on the continuum. The clear isolation of surface forces of stress in a continuum is usually attributed to Cauchy.

Later, we shall examine stresses more generally, but for now let us define an ***ideal fluid*** as one with the following property: *For any motion of the fluid there is a function $p(\mathbf{x}, t)$ called the* ***pressure*** *such that if S is a surface in the fluid with a chosen unit normal $\mathbf{n}$, the force of stress exerted across the surface S per unit area at $\mathbf{x} \in S$ at time t is $p(\mathbf{x}, t)\mathbf{n}$; i.e.,*

$$\text{force across } S \text{ per unit area} = p(\mathbf{x}, t)\mathbf{n}.$$

Note that the force is in the direction $\mathbf{n}$ and that the force acts orthogonally to the surface S; that is, there are no tangential forces (see Figure 1.1.3).

Of course, the concept of an ideal fluid as a mathematical definition is not subject to dispute. However, the physical relevance of the notion (or mathematical theorems we deduce from it) must be checked by experiment. As we shall see later, ideal fluids exclude many interesting real physical phenomena, but nevertheless form a crucial component of a more complete theory.

Intuitively, the absence of tangential forces implies that there is no way for rotation to start in a fluid, nor, if it is there at the beginning, to stop. This idea will be amplified in the next section. However, even here we can detect physical trouble for ideal fluids because of the abundance of rotation in real fluids (near the oars of a rowboat, in tornadoes, etc.).

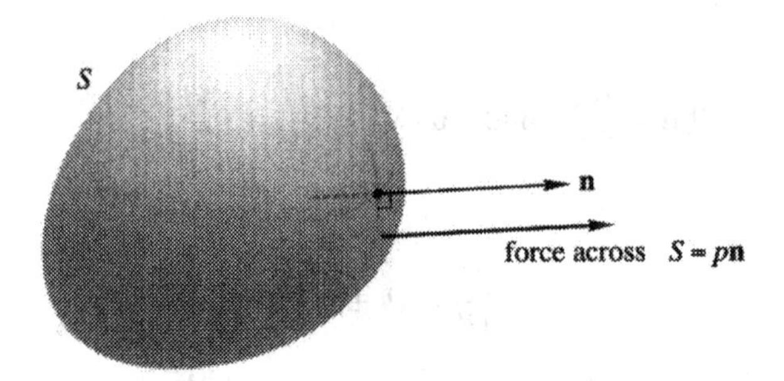

FIGURE 1.1.3. Pressure forces across a surface S.

If W is a region in the fluid at a particular instant of time t, the total force exerted *on* the fluid inside W by means of stress on its boundary is

$$\mathbf{S}_{\partial W} = \{\text{force on } W\} = -\int_{\partial W} p\mathbf{n}\, dA$$

(negative because $\mathbf{n}$ points outward). If $\mathbf{e}$ is any fixed vector in space, the divergence theorem gives

$$\mathbf{e} \cdot \mathbf{S}_{\partial W} = -\int_{\partial W} p\mathbf{e} \cdot \mathbf{n}\, dA = -\int_W \operatorname{div}(p\mathbf{e})\, dV = -\int_W (\operatorname{grad} p) \cdot \mathbf{e}\, dV.$$

Thus,

$$\mathbf{S}_{\partial W} = -\int_W \operatorname{grad} p\, dV.$$

If $\mathbf{b}(\mathbf{x}, t)$ denotes the given body force *per unit mass*, the total body force is

$$\mathbf{B} = \int_W \rho \mathbf{b}\, dV.$$

Thus, on any piece of fluid material,

$$\text{force per unit volume} = -\operatorname{grad} p + \rho \mathbf{b}.$$

By Newton's second law (force = mass × acceleration) we are led to the differential form of the law of ***balance of momentum***:

$$\boxed{\rho \frac{D\mathbf{u}}{Dt} = -\operatorname{grad} p + \rho \mathbf{b}.} \tag{BM1}$$

Next we shall derive an integral form of balance of momentum in two ways. We derive it first as a deduction from the differential form and second from basic principles.

From balance of momentum in differential form, we have

$$\rho \frac{\partial \mathbf{u}}{\partial t} = -\rho(\mathbf{u} \cdot \nabla)\mathbf{u} - \nabla p + \rho \mathbf{b}$$

and so, using the equation of continuity,

$$\frac{\partial}{\partial t}(\rho \mathbf{u}) = -\operatorname{div}(\rho \mathbf{u})\mathbf{u} - \rho(\mathbf{u} \cdot \nabla)\mathbf{u} - \nabla p + \rho \mathbf{b}.$$

If $\mathbf{e}$ is any fixed vector in space, one checks that

$$\begin{aligned} \mathbf{e} \cdot \frac{\partial}{\partial t}(\rho \mathbf{u}) &= -\operatorname{div}(\rho \mathbf{u})\mathbf{u} \cdot \mathbf{e} - \rho(\mathbf{u} \cdot \nabla)\mathbf{u} \cdot \mathbf{e} - (\nabla p) \cdot \mathbf{e} + \rho \mathbf{b} \cdot \mathbf{e} \\ &= -\operatorname{div}(p\mathbf{e} + \rho \mathbf{u}(\mathbf{u} \cdot \mathbf{e})) + \rho \mathbf{b} \cdot \mathbf{e}. \end{aligned}$$

Therefore, if W is a *fixed* volume in space, the rate of change of momentum in direction $\mathbf{e}$ in W is

$$\mathbf{e} \cdot \frac{d}{dt} \int_W \rho \mathbf{u}\, dV = -\int_{\partial W} (p\mathbf{e} + \rho \mathbf{u}(\mathbf{e} \cdot \mathbf{u})) \cdot \mathbf{n}\, dA + \int_W \rho \mathbf{b} \cdot \mathbf{e}\, dV$$

by the divergence theorem. Thus, the integral form of balance of momentum becomes:

$$\frac{d}{dt} \int_W \rho \mathbf{u}\, dV = -\int_{\partial W} (p\mathbf{n} + \rho \mathbf{u}(\mathbf{u} \cdot \mathbf{n}))\, dA + \int_W \rho \mathbf{b}\, dV. \qquad \text{(BM2)}$$

The quantity $p\mathbf{n}+\rho\mathbf{u}(\mathbf{u}\cdot\mathbf{n})$ is the ***momentum flux per unit area*** crossing ∂W, where $\mathbf{n}$ is the unit outward normal to ∂W.

This derivation of the integral balance law for momentum proceeded via the differential law. With an eye to assuming as little differentiability as possible, it is useful to proceed to the integral law directly and, as with conservation of mass, derive the differential form from it. To do this carefully requires us to introduce some useful notions.

As earlier, let D denote the region in which the fluid is moving. Let $\mathbf{x} \in D$ and let us write $\varphi(\mathbf{x}, t)$ for the trajectory followed by the particle that is at point $\mathbf{x}$ at time $t = 0$. We will assume φ is smooth enough so the following manipulations are legitimate and for fixed t, φ is an invertible mapping. Let φ_t denote the map $\mathbf{x} \mapsto \varphi(\mathbf{x}, t)$; that is, with fixed t, this map advances each fluid particle from its position at time $t = 0$ to its position at time t. Here, of course, the subscript does *not* denote differentiation. We call φ the ***fluid flow map***. If W is a region in D, then $\varphi_t(W) = W_t$ is the volume W *moving with the fluid*. See Figure 1.1.4.

The "primitive" integral form of balance of momentum states that

$$\frac{d}{dt} \int_{W_t} \rho \mathbf{u}\, dV = S_{\partial W_t} + \int_{W_t} \rho \mathbf{b}\, dV, \qquad \text{(BM3)}$$

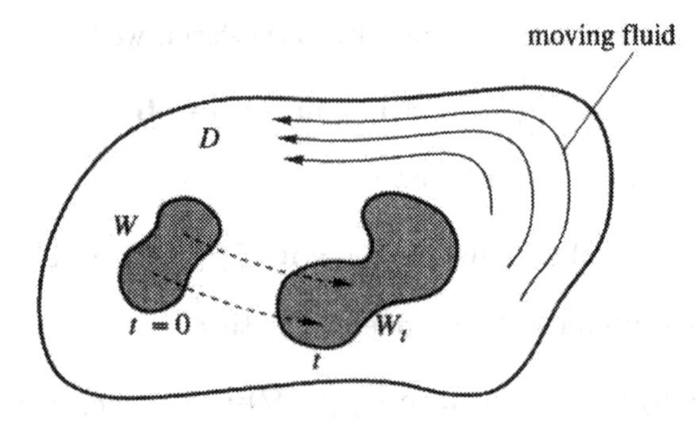

FIGURE 1.1.4. W_t is the image of W as particles of fluid in W flow for time t.

that is, the rate of change of momentum of a moving piece of fluid equals the total force (surface stresses plus body forces) acting on it.

These *two forms of balance of momentum* (BM1) and (BM3) *are equivalent*. To prove this, we use the change of variables theorem to write

$$\frac{d}{dt}\int_{W_t} \rho\mathbf{u}\, dV = \frac{d}{dt}\int_{W} (\rho\mathbf{u})(\varphi(\mathbf{x},t),t)J(\mathbf{x},t)\, dV,$$

where $J(\mathbf{x},t)$ is the Jacobian determinant of the map φ_t. Because the volume is fixed at its initial position, we may differentiate under the integral sign. Note that

$$\frac{\partial}{\partial t}(\rho\mathbf{u})(\varphi(\mathbf{x},t),t) = \left(\frac{D}{Dt}\rho\mathbf{u}\right)(\varphi(\mathbf{x},t),t)$$

is the material derivative, as was shown earlier. (If you prefer, this equality says that D/Dt is differentiation following the fluid.) Next, we learn how to differentiate $J(\mathbf{x},t)$.

Lemma

$$\frac{\partial}{\partial t}J(\mathbf{x},t) = J(\mathbf{x},t)[\operatorname{div}\mathbf{u}(\varphi(\mathbf{x},t),t)].$$

Proof Write the components of φ as $\xi(\mathbf{x},t), \eta(\mathbf{x},t)$, and $\zeta(\mathbf{x},t)$. First, observe that

$$\frac{\partial}{\partial t}\varphi(\mathbf{x},t) = \mathbf{u}(\varphi(\mathbf{x},t),t),$$

by definition of the velocity field of the fluid.

The determinant J can be differentiated by recalling that the determinant of a matrix is multilinear in the columns (or rows). Thus, holding $\mathbf{x}$

fixed throughout, we have

$$\frac{\partial}{\partial t}J = \begin{bmatrix} \frac{\partial}{\partial t}\frac{\partial \xi}{\partial x} & \frac{\partial \eta}{\partial x} & \frac{\partial \zeta}{\partial x} \\ \frac{\partial}{\partial t}\frac{\partial \xi}{\partial y} & \frac{\partial \eta}{\partial y} & \frac{\partial \zeta}{\partial y} \\ \frac{\partial}{\partial t}\frac{\partial \xi}{\partial z} & \frac{\partial \eta}{\partial z} & \frac{\partial \zeta}{\partial z} \end{bmatrix} + \begin{bmatrix} \frac{\partial \xi}{\partial x} & \frac{\partial}{\partial t}\frac{\partial \eta}{\partial x} & \frac{\partial \zeta}{\partial x} \\ \frac{\partial \xi}{\partial y} & \frac{\partial}{\partial t}\frac{\partial \eta}{\partial y} & \frac{\partial \zeta}{\partial y} \\ \frac{\partial \xi}{\partial z} & \frac{\partial}{\partial t}\frac{\partial \eta}{\partial z} & \frac{\partial \zeta}{\partial z} \end{bmatrix}$$

$$+ \begin{bmatrix} \frac{\partial \xi}{\partial x} & \frac{\partial \eta}{\partial x} & \frac{\partial}{\partial t}\frac{\partial \zeta}{\partial x} \\ \frac{\partial \xi}{\partial y} & \frac{\partial \eta}{\partial y} & \frac{\partial}{\partial t}\frac{\partial \zeta}{\partial y} \\ \frac{\partial \xi}{\partial z} & \frac{\partial \eta}{\partial z} & \frac{\partial}{\partial t}\frac{\partial \zeta}{\partial z} \end{bmatrix}.$$

Now write

$$\frac{\partial}{\partial t}\frac{\partial \xi}{\partial x} = \frac{\partial}{\partial x}\frac{\partial \xi}{\partial t} = \frac{\partial}{\partial x}u(\varphi(\mathbf{x},t),t),$$

$$\frac{\partial}{\partial t}\frac{\partial \xi}{\partial y} = \frac{\partial}{\partial y}\frac{\partial \xi}{\partial t} = \frac{\partial}{\partial y}u(\varphi(\mathbf{x},t),t),$$

$$\vdots$$

$$\frac{\partial}{\partial t}\frac{\partial \zeta}{\partial z} = \frac{\partial}{\partial z}\frac{\partial \zeta}{\partial t} = \frac{\partial}{\partial z}w(\varphi(\mathbf{x},t),t).$$

The components u, v, and w of $\mathbf{u}$ in this expression are functions of x, y, and z through $\varphi(\mathbf{x}, t)$; therefore,

$$\frac{\partial}{\partial x}u(\varphi(\mathbf{x},t),t) = \frac{\partial u}{\partial \xi}\frac{\partial \xi}{\partial x} + \frac{\partial u}{\partial \eta}\frac{\partial \eta}{\partial x} + \frac{\partial u}{\partial \zeta}\frac{\partial \zeta}{\partial x},$$

$$\vdots$$

$$\frac{\partial}{\partial z}w(\varphi(\mathbf{x},t),t) = \frac{\partial w}{\partial \xi}\frac{\partial \xi}{\partial z} + \frac{\partial w}{\partial \eta}\frac{\partial \eta}{\partial z} + \frac{\partial w}{\partial \zeta}\frac{\partial \zeta}{\partial z}.$$

When these are substituted into the above expression for $\partial J/\partial t$, one gets for the respective terms

$$\frac{\partial u}{\partial x}J + \frac{\partial v}{\partial y}J + \frac{\partial w}{\partial z}J = (\operatorname{div}\mathbf{u})J. \qquad \blacksquare$$

From this lemma, we get

$$\frac{d}{dt}\int_{W_t} \rho\mathbf{u}\,dV = \int_W \left\{\left(\frac{D}{Dt}\rho\mathbf{u}\right)(\varphi(\mathbf{x},t),t) + (\rho\mathbf{u})(\operatorname{div}\mathbf{u})(\varphi(\mathbf{x},t),t)\right\} \times J(\mathbf{x},t)\,dV$$

$$= \int_{W_t} \left\{\frac{D}{Dt}(\rho\mathbf{u}) + (\rho\ \operatorname{div}\mathbf{u})\mathbf{u}\right\} dV,$$

where the change of variables theorem was again used. By conservation of mass,

$$\frac{D}{Dt}\rho + \rho\ \operatorname{div}\mathbf{u} = \frac{\partial\rho}{\partial t} + \operatorname{div}(\rho\mathbf{u}) = 0,$$

and thus

$$\frac{d}{dt}\int_{W_t} \rho\mathbf{u}\,dV = \int_{W_t} \rho\frac{D\mathbf{u}}{Dt}\,dV.$$

In fact, this argument proves the following theorem.

Transport Theorem *For any function f of $\mathbf{x}$ and t, we have*

$$\boxed{\frac{d}{dt}\int_{W_t} \rho f\,dV = \int_{W_t} \rho\frac{Df}{Dt}\,dV.}$$

In a similar way, one can derive a form of the transport theorem *without a mass density* factor included, namely,

$$\boxed{\frac{d}{dt}\int_{W_t} f\,dV = \int_{W_t} \left(\frac{\partial f}{\partial t} + \operatorname{div}(f\mathbf{u})\right) dV.}$$

If W, and hence, W_t, is arbitrary and the integrands are continuous, we have proved that the "primitive" integral form of balance of momentum is equivalent to the differential form (BM1). Hence, all three forms of balance of momentum—(BM1), (BM2), and (BM3)—are mutually equivalent. As an exercise, the reader should derive the two integral forms of balance of momentum directly from each other.

The lemma $\partial J/\partial t = (\operatorname{div}\mathbf{u})\,J$ is also useful in understanding incompressibility. In terms of the notation introduced earlier, we call a flow ***incompressible*** if for any fluid subregion W,

$$\text{volume}(W_t) = \int_{W_t} dV = \text{constant in } t.$$

Thus, incompressibility is equivalent to

$$0 = \frac{d}{dt}\int_{W_t} dV = \frac{d}{dt}\int_W J\,dV = \int_W (\operatorname{div}\mathbf{u})J\,dV = \int_{W_t} (\operatorname{div}\mathbf{u})\,dV$$

for all moving regions W_t. Thus, the following are equivalent:

(i) *the fluid is incompressible*;

(ii) $\operatorname{div}\mathbf{u} = 0$;

(iii) $J \equiv 1$.

From the equation of continuity

$$\frac{\partial\rho}{\partial t} + \operatorname{div}(\rho\mathbf{u}) = 0, \quad \text{i.e.,} \quad \frac{D\rho}{Dt} + \rho\operatorname{div}\mathbf{u} = 0,$$

and the fact that $\rho > 0$, we see that a *fluid is incompressible if and only if* $D\rho/Dt = 0$, *that is, the mass density is constant following the fluid.* If the fluid is ***homogeneous***, that is, $\rho =$ constant in space, it also follows that the flow is incompressible if and only if ρ is constant in time. Problems involving inhomogeneous incompressible flow occur, for example, in oceanography.

We shall now "solve" the equation of continuity by expressing ρ in terms of its value at $t = 0$, the flow map $\varphi(\mathbf{x}, t)$, and its Jacobian $J(\mathbf{x}, t)$. Indeed, set $f = 1$ in the transport theorem and conclude the equivalent condition for mass conservation,

$$\frac{d}{dt}\int_{W_t} \rho\,dV = 0$$

and thus,

$$\int_{W_t} \rho(\mathbf{x}, t)dV = \int_{W_0} \rho(\mathbf{x}, 0)\,dV.$$

Changing variables, we obtain

$$\int_{W_0} \rho(\varphi(\mathbf{x}, t), t)J(\mathbf{x}, t)\,dV = \int_{W_0} \rho(\mathbf{x}, 0)\,dV.$$

Because W_0 is arbitrary, we get

$$\boxed{\rho(\varphi(\mathbf{x}, t), t)J(\mathbf{x}, t) = \rho(\mathbf{x}, 0)}$$

as another form of mass conservation. As a corollary, a fluid that is homogeneous at $t = 0$ but is compressible will generally not remain homogeneous. However, the fluid will remain homogeneous if it is incompressible. The example $\varphi((x, y, z), t) = ((1 + t)x, y, z)$ has $J((x, y, z), t) = 1 + t$ so the flow is not incompressible, yet for $\rho((x, y, z), t) = 1/(1 + t)$, one has mass conservation and homogeneity for all time.

iii Conservation of Energy

So far we have developed the equations

$$\rho\frac{D\mathbf{u}}{Dt} = -\operatorname{grad} p + \rho\mathbf{b} \quad \text{(balance of momentum)}$$

and

$$\frac{D\rho}{Dt} + \rho \operatorname{div} \mathbf{u} = 0 \qquad \text{(conservation of mass).}$$

These are four equations if we work in 3-dimensional space (or $n+1$ equations if we work in n-dimensional space), because the equation for $D\mathbf{u}/Dt$ is a vector equation composed of three scalar equations. However, we have *five* functions: $\mathbf{u}$, ρ, and p. Thus, one might suspect that to specify the fluid motion completely, one more equation is needed. This is in fact true, and conservation of energy will supply the necessary equation in fluid mechanics. This situation is more complicated for general continua, and issues of general thermodynamics would need to be discussed for a complete treatment. We shall confine ourselves to two special cases here, and later we shall treat another case for an ideal gas.

For fluid moving in a domain D, with velocity field $\mathbf{u}$, the ***kinetic energy*** contained in a region $W \subset D$ is

$$E_{\text{kinetic}} = \frac{1}{2}\int_W \rho\|\mathbf{u}\|^2\, dV$$

where $\|\mathbf{u}\|^2 = (u^2 + v^2 + w^2)$ is the square length of the vector function $\mathbf{u}$. We assume that the total energy of the fluid can be written as

$$E_{\text{total}} = E_{\text{kinetic}} + E_{\text{internal}}$$

where E_{internal} is the ***internal energy***, which is energy we cannot "see" on a macroscopic scale, and derives from sources such as intermolecular potentials and internal molecular vibrations. If energy is pumped into the fluid or if we allow the fluid to do work, E_{total} will change.

The rate of change of kinetic energy of a moving portion W_t of fluid is calculated using the transport theorem as follows:

$$\begin{aligned}
\frac{d}{dt}E_{\text{kinetic}} &= \frac{d}{dt}\left[\frac{1}{2}\int_{W_t}\rho\|\mathbf{u}\|^2\, dV\right]\\
&= \frac{1}{2}\int_{W_t}\rho\frac{D\|\mathbf{u}\|^2}{Dt}\, dV\\
&= \int_{W_t}\rho\left(\mathbf{u}\cdot\left(\frac{\partial\mathbf{u}}{\partial t} + (\mathbf{u}\cdot\nabla)\mathbf{u}\right)\right) dV.
\end{aligned}$$

Here we have used the following Euclidean coordinate calculation

$$\begin{aligned}\frac{1}{2}\frac{D}{Dt}\|\mathbf{u}\|^2 &= \frac{1}{2}\frac{\partial}{\partial t}(u^2+v^2+w^2) + \frac{1}{2}\left(u\frac{\partial}{\partial x}(u^2+v^2+w^2)\right.\\ &\quad \left. + v\frac{\partial}{\partial y}(u^2+v^2+w^2) + w\frac{\partial}{\partial z}(u^2+v^2+w^2)\right)\\ &= u\frac{\partial u}{\partial t} + v\frac{\partial v}{\partial t} + w\frac{\partial w}{\partial t} + u\left(u\frac{\partial u}{\partial x} + v\frac{\partial v}{\partial x} + w\frac{\partial w}{\partial x}\right)\\ &\quad + v\left(u\frac{\partial u}{\partial y} + v\frac{\partial v}{\partial y} + w\frac{\partial w}{\partial y}\right) + w\left(u\frac{\partial u}{\partial z} + v\frac{\partial v}{\partial z} + w\frac{\partial w}{\partial z}\right)\\ &= \mathbf{u}\cdot\frac{\partial \mathbf{u}}{\partial t} + \mathbf{u}\cdot(\mathbf{u}\cdot\nabla)\mathbf{u}.\end{aligned}$$

A general discussion of energy conservation requires more thermodynamics than we shall need. We limit ourselves here to two examples of energy conservation; a third will be given in Chapter **3**.

1 Incompressible Flows

Here we assume *all* the energy is kinetic and that *the rate of change of kinetic energy in a portion of fluid equals the rate at which the pressure and body forces do work:*

$$\frac{d}{dt}E_{\text{kinetic}} = -\int_{\partial W_t} p\mathbf{u}\cdot\mathbf{n}\,dA + \int_{W_t}\rho\mathbf{u}\cdot\mathbf{b}\,dV.$$

By the divergence theorem and our previous formulas, this becomes

$$\begin{aligned}\int_{W_t}\rho\left\{\mathbf{u}\cdot\left(\frac{\partial\mathbf{u}}{\partial t} + \mathbf{u}\cdot\nabla\mathbf{u}\right)\right\}dV &= -\int_{W_t}(\operatorname{div}(p\mathbf{u}) - \rho\mathbf{u}\cdot\mathbf{b})\,dV\\ &= -\int_{W_t}(\mathbf{u}\cdot\nabla p - \rho\mathbf{u}\cdot\mathbf{b})\,dV\end{aligned}$$

because $\operatorname{div}\mathbf{u} = 0$. The preceding equation is also a consequence of balance of momentum. This argument, in addition, shows that *if we assume* $E = E_{\text{kinetic}}$, *then the fluid must be incompressible* (unless $p = 0$). In summary, in this incompressible case, the ***Euler equations*** are:

$$\begin{aligned}\rho\frac{D\mathbf{u}}{Dt} &= -\operatorname{grad}p + \rho\mathbf{b}\\ \frac{D\rho}{Dt} &= 0\\ \operatorname{div}\mathbf{u} &= 0\end{aligned}$$

with the boundary conditions

$$\mathbf{u}\cdot\mathbf{n} = 0 \quad \text{on } \partial D.$$

2 Isentropic Fluids

A compressible flow will be called ***isentropic*** if there is a function w, called the ***enthalpy***, such that

$$\operatorname{grad} w = \frac{1}{\rho} \operatorname{grad} p.$$

This terminology comes from thermodynamics. We shall not need a detailed discussion of thermodynamics concepts in this book, and so it is omitted, except for a brief discussion of entropy in Chapter **3** in the context of ideal gases. For the readers' convenience, we just make a few general comments.

In thermodynamics one has the following basic quantities, each of which is a function of $\mathbf{x}, t$ depending on a given flow:

$p =$ ***pressure***
$\rho =$ ***density***
$T =$ ***temperature***
$s =$ ***entropy***
$w =$ ***enthalpy*** (per unit mass)
$\epsilon = w - (p/\rho) =$ ***internal energy*** (per unit mass).

These quantities are related by the ***First Law of Thermodynamics***, which we accept as a basic principle:[2]

$$dw = T\,ds + \frac{1}{\rho}\,dp \qquad \text{(TD1)}$$

The first law is a statement of conservation of energy; a statement equivalent to (TD1) is, as is readily verified,

$$d\epsilon = T\,ds + \frac{p}{\rho^2}\,d\rho. \qquad \text{(TD2)}$$

If the pressure is a function of ρ only, then the flow is clearly isentropic with s as a constant (hence the name ***isentropic***) and

$$w = \int^{\rho} \frac{p'(\lambda)}{\lambda}\,d\lambda,$$

which is the integrated version of $dw = dp/\rho$ (see TD1). As above, the internal energy $\epsilon = w - (p/\rho)$ then satisfies $d\epsilon = (p d\rho)/\rho^2$ (see TD2) or, as a function of ρ,

$$p = \rho^2 \frac{\partial \varepsilon}{\partial \rho}, \quad \text{or} \quad \epsilon = \int^{\rho} \frac{p(\lambda)}{\lambda^2}\,d\lambda.$$

---

[2] A. Sommerfeld [1964] *Thermodynamics and Statistical Mechanics*, reprinted by Academic Press, Chapters **1** and **4**.

For isentropic flows with p a function of ρ, the integral form of energy balance reads as follows: *The rate of change of energy in a portion of fluid equals the rate at which work is done on it:*

$$\begin{aligned}\frac{d}{dt}E_{\text{total}} &= \frac{d}{dt}\int_{W_t}\left(\tfrac{1}{2}\rho\|\mathbf{u}\|^2 + \rho\epsilon\right)dV \\ &= \int_{W_t}\rho\mathbf{u}\cdot\mathbf{b}\,dV - \int_{\partial W_t} p\mathbf{u}\cdot\mathbf{n}\,dA.\end{aligned} \tag{BE}$$

This follows from balance of momentum using our earlier expression for $(d/dt)E_{\text{kinetic}}$, the transport theorem, and $p = \rho^2\partial\epsilon/\partial\rho$. Alternatively, one can start with the assumption that p is a function of ρ and then (BE) and balance of mass and momentum implies that $p = \rho^2\partial\epsilon/\partial\rho$, which is *equivalent* to $dw = dp/\rho$, as we have seen.[3]

Euler's equations for isentropic flow are thus

$$\begin{aligned}\frac{\partial\mathbf{u}}{\partial t} + (\mathbf{u}\cdot\nabla)\mathbf{u} &= -\nabla w + \mathbf{b}, \\ \frac{\partial\rho}{\partial t} + \operatorname{div}(\rho\mathbf{u}) &= 0\end{aligned}$$

in D, and

$$\mathbf{u}\cdot\mathbf{n} = 0$$

on ∂D (or $\mathbf{u}\cdot\mathbf{n} = \mathbf{V}\cdot\mathbf{n}$ if ∂D is moving with velocity $\mathbf{V}$).

Later, we will see that in general these equations lead to a well-posed initial value problem only if $p'(\rho) > 0$. This agrees with the common experience that increasing the surrounding pressure on a volume of fluid causes a decrease in occupied volume and hence an increase in density.

Gases can often be viewed as isentropic, with

$$p = A\rho^\gamma,$$

where A and γ are constants and $\gamma \geq 1$. Here,

$$w = \int^{\rho}\frac{\gamma A s^{\gamma-1}}{s}\,ds = \frac{\gamma A\rho^{\gamma-1}}{\gamma-1} \quad\text{and}\quad \epsilon = \frac{A\rho^{\gamma-1}}{\gamma-1}.$$

Cases 1 and 2 above are rather opposite. For instance, if $\rho = \rho_0$ is a constant for an incompressible fluid, then clearly p cannot be an invertible function of ρ. However, the case $\rho =$ constant may be regarded as a limiting case $p'(\rho) \to \infty$. In case 2, p is an explicit function of ρ (and therefore

[3]One can carry this even further and use balance of energy and its invariance under Euclidean motions to derive balance of momentum and mass, a result of Green and Naghdi. See Marsden and Hughes [1994] for a proof and extensions of the result that include formulas such as $p = p^2\partial\varepsilon/\partial p$ amongst the consequences as well.

depends on $\mathbf{u}$ through the coupling of ρ and $\mathbf{u}$ in the equation of continuity); in case 1, p is implicitly determined by the condition $\operatorname{div} \mathbf{u} = 0$. We shall discuss these points again later.

Finally, notice that in neither case 1 or 2 is the possibility of a loss of kinetic energy due to friction taken into account. This will be discussed at length in §**1.3**.

Given a fluid flow with velocity field $\mathbf{u}(\mathbf{x}, t)$, a ***streamline*** at a fixed time is an integral curve of $\mathbf{u}$; that is, if $\mathbf{x}(s)$ is a streamline at the instant t, it is a curve parametrized by a variable, say s, that satisfies

$$\frac{d\mathbf{x}}{ds} = \mathbf{u}(\mathbf{x}(s), t), \qquad t \text{ fixed.}$$

We define a fixed ***trajectory*** to be the curve traced out by a particle as time progresses, as explained at the beginning of this section. Thus, a trajectory is a solution of the differential equation

$$\frac{d\mathbf{x}}{dt} = \mathbf{u}(\mathbf{x}(t), t)$$

with suitable initial conditions. If $\mathbf{u}$ is independent of t (i.e., $\partial_t \mathbf{u} = 0$), streamlines and trajectories coincide. In this case, the flow is called ***stationary***.

Bernoulli's Theorem *In stationary isentropic flows and in the absence of external forces, the quantity*

$$\tfrac{1}{2}\|\mathbf{u}\|^2 + w$$

is constant along streamlines. The same holds for homogeneous ($\rho =$ constant in space $= \rho_0$) incompressible flow with w replaced by p/ρ_0. The conclusions remain true if a force $\mathbf{b}$ is present and is conservative; i.e., $\mathbf{b} = -\nabla\varphi$ for some function φ, with w replaced by $w + \varphi$.

Proof From the table of vector identities at the back of the book, one has

$$\tfrac{1}{2}\nabla(\|\mathbf{u}\|^2) = (\mathbf{u} \cdot \nabla)\mathbf{u} + \mathbf{u} \times (\nabla \times \mathbf{u}).$$

Because the flow is steady, the equations of motion give

$$(\mathbf{u} \cdot \nabla)\mathbf{u} = -\nabla w$$

and so

$$\nabla\left(\tfrac{1}{2}\|\mathbf{u}\|^2 + w\right) = \mathbf{u} \times (\nabla \times \mathbf{u}).$$

Let $\mathbf{x}(s)$ be a streamline. Then

$$\frac{1}{2}\left(\|\mathbf{u}\|^2 + w\right)\Big|_{\mathbf{x}(s_1)}^{\mathbf{x}(s_2)} = \int_{\mathbf{x}(s_1)}^{\mathbf{x}(s_2)} \nabla\left(\tfrac{1}{2}\|\mathbf{u}\|^2 + w\right)\cdot \mathbf{x}'(s)\,ds$$
$$= \int_{\mathbf{x}(s_1)}^{\mathbf{x}(s_2)} (\mathbf{u}\times(\nabla\times\mathbf{u}))\cdot\mathbf{x}'(s)\,ds = 0$$

because $\mathbf{x}'(s) = \mathbf{u}(\mathbf{x}(s))$ is orthogonal to $\mathbf{u}\times(\nabla\times\mathbf{u})$. ■

See Exercise 1.1-3 at the end of this section for another view of why the combination $\frac{1}{2}\|\mathbf{u}\|^2 + w$ is the correct quantity in Bernoulli's theorem.

We conclude this section with an example that shows the limitations of the assumptions we have made so far.

Example Consider a fluid-filled channel, as in Figure 1.1.5.

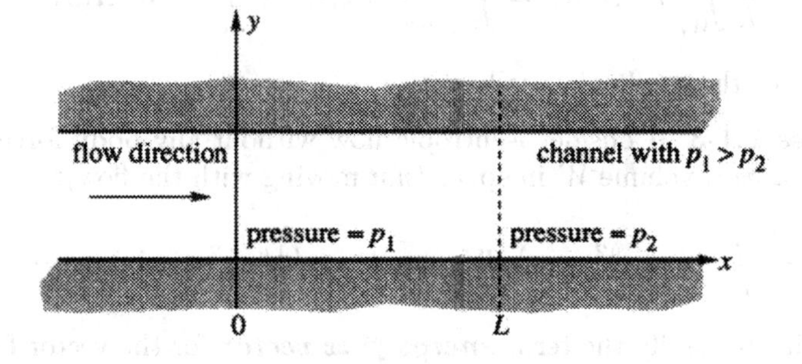

FIGURE 1.1.5. Fluid flow in a channel.

Suppose that the pressure p_1 at $x = 0$ is larger than that at $x = L$ so the fluid is pushed from left to right. We seek a solution of Euler's incompressible homogeneous equations in the form

$$\mathbf{u}(x,y,t) = (u(x,t),0) \quad \text{and} \quad p(x,y,t) = p(x).$$

Incompressibility implies $\partial_x u = 0$. Thus, Euler's equations become $\rho_0 \partial_t u = -\partial_x p$. This implies that $\partial_x^2 p = 0$, and so

$$p(x) = p_1 - \left(\frac{p_1 - p_2}{L}\right)x.$$

Substitution into $\rho_0\partial_t u = -\partial_x p$ and integration yields

$$u = \frac{p_1 - p_2}{\rho_0 L}t + \text{constant}.$$

This solution suggests that the velocity in channel flow with a constant pressure gradient increases indefinitely. Of course, this cannot be the case in a real flow; however, in our modeling, we have not yet taken friction into account. The situation will be remedied in **§1.3.** ♦

Exercises

◇ **Exercise 1.1-1** Prove the following properties of the material derivative

(i) $\frac{D}{Dt}(f+g) = \frac{Df}{Dt} + \frac{Dg}{Dt}$,

(ii) $\frac{D}{Dt}(f \cdot g) = f\frac{Dg}{Dt} + g\frac{Df}{Dt}$ (Leibniz or product rule),

(iii) $\frac{D}{Dt}(h \circ g) = (h' \circ g)\frac{Dg}{Dt}$ (chain rule).

◇ **Exercise 1.1-2** Use the transport theorem to establish the following formula of Reynolds:

$$\frac{d}{dt}\int_{W_t} f(x,t)\, dV = \int_{W_t} \frac{\partial f}{\partial t}(x,t)\, dV + \int_{\partial W_t} f\mathbf{u} \cdot \mathbf{n}\, dA.$$

Interpret the result physically.

◇ **Exercise 1.1-3** Consider isentropic flow without any body force. Show that for a *fixed* volume W in space (*not* moving with the flow).

$$\frac{d}{dt}\int_W \left(\tfrac{1}{2}\rho\|\mathbf{u}\|^2 + \rho\epsilon\right) dV = -\int_{\partial W} \rho\left(\tfrac{1}{2}\|\mathbf{u}\|^2 + w\right)\mathbf{u} \cdot \mathbf{n}\, dA.$$

Use this to justify the term ***energy flux vector*** for the vector function $\rho\mathbf{u}\left(\frac{1}{2}\|\mathbf{u}\|^2 + w\right)$ and compare with Bernoulli's theorem.

1.2 Rotation and Vorticity

If the velocity field of a fluid is $\mathbf{u} = (u, v, w)$, then its curl,

$$\boldsymbol{\xi} = \nabla \times \mathbf{u} = (\partial_y w - \partial_z v, \partial_z u - \partial_x w, \partial_x v - \partial_y u)$$

is called the ***vorticity field*** of the flow.

We shall now demonstrate that in a small neighborhood of each point of the fluid, $\mathbf{u}$ *is the sum of a (rigid) translation, a deformation (defined later), and a (rigid) rotation with rotation vector* $\boldsymbol{\xi}/2$. This is in fact a general statement about vector fields $\mathbf{u}$ on $\mathbb{R}^3$; the specific features of fluid mechanics are irrelevant for this discussion. Let $\mathbf{x}$ be a point in $\mathbb{R}^3$, and let $\mathbf{y} = \mathbf{x} + \mathbf{h}$ be a nearby point. What we shall prove is that

$$\mathbf{u}(\mathbf{y}) = \mathbf{u}(\mathbf{x}) + \mathbf{D}(\mathbf{x}) \cdot \mathbf{h} + \tfrac{1}{2}\boldsymbol{\xi}(\mathbf{x}) \times \mathbf{h} + O(h^2), \tag{1.2.1}$$

where $\mathbf{D}(x)$ is a symmetric 3×3 matrix and $h^2 = \|\mathbf{h}\|^2$ is the squared length of $\mathbf{h}$. We shall discuss the meaning of the several terms later.

Proof of Formula (1.2.1) Let

$$\nabla \mathbf{u} = \begin{bmatrix} \partial_x u & \partial_y u & \partial_z u \\ \partial_x v & \partial_y v & \partial_z v \\ \partial_x w & \partial_y w & \partial_z w \end{bmatrix}$$

denote the Jacobian matrix of $\mathbf{u}$. By Taylor's theorem,

$$\mathbf{u}(\mathbf{y}) = \mathbf{u}(\mathbf{x}) + \nabla \mathbf{u}(\mathbf{x}) \cdot \mathbf{h} + O(h^2), \tag{1.2.2}$$

where $\nabla \mathbf{u}(\mathbf{x}) \cdot \mathbf{h}$ is a matrix multiplication, with $\mathbf{h}$ regarded as a column vector. Let

$$\mathbf{D} = \tfrac{1}{2}\left[\nabla \mathbf{u} + (\nabla \mathbf{u})^T\right],$$

where ^{T} denotes the transpose, and

$$\mathbf{S} = \tfrac{1}{2}\left[\nabla \mathbf{u} - (\nabla \mathbf{u})^T\right].$$

Thus,

$$\nabla \mathbf{u} = \mathbf{D} + \mathbf{S}. \tag{1.2.3}$$

It is easy to check that the coordinate expression for $\mathbf{S}$ is

$$\mathbf{S} = \frac{1}{2}\begin{bmatrix} 0 & -\xi_3 & \xi_2 \\ \xi_3 & 0 & -\xi_1 \\ -\xi_2 & \xi_1 & 0 \end{bmatrix}$$

and that

$$\mathbf{S} \cdot \mathbf{h} = \tfrac{1}{2}\boldsymbol{\xi} \times \mathbf{h}, \tag{1.2.4}$$

where $\boldsymbol{\xi} = (\xi_1, \xi_2, \xi_3)$. Substitution of (1.2.3) and (1.2.4) into (1.2.2) yields (1.2.1). ∎

Because $\mathbf{D}$ is a symmetric matrix,

$$\mathbf{D}(\mathbf{x}) \cdot \mathbf{h} = \operatorname{grad}_h \psi(\mathbf{x}, \mathbf{h}),$$

where ψ is the quadratic form associated with $\mathbf{D}$; *i.e.*,

$$\psi(\mathbf{x}, \mathbf{h}) = \tfrac{1}{2}\langle \mathbf{D}(\mathbf{x}) \cdot \mathbf{h}, \mathbf{h}\rangle,$$

where $\langle\, ,\rangle$ is the inner product of $\mathbb{R}^3$. We call $\mathbf{D}$ the ***deformation tensor***. We now discuss its physical interpretation. Because $\mathbf{D}$ is symmetric, there is, for $\mathbf{x}$ fixed, an orthonormal basis $\tilde{\mathbf{e}}_1, \tilde{\mathbf{e}}_2, \tilde{\mathbf{e}}_3$ in which $\mathbf{D}$ is diagonal:

$$\mathbf{D} = \begin{bmatrix} d_1 & 0 & 0 \\ 0 & d_2 & 0 \\ 0 & 0 & d_3 \end{bmatrix}.$$

Keep $\mathbf{x}$ fixed and consider the original vector field as a function of y. The motion of the fluid is described by the equations

$$\frac{d\mathbf{y}}{dt} = \mathbf{u}(\mathbf{y}).$$

If we ignore all terms in (1.2.1) except $\mathbf{D} \cdot \mathbf{h}$, we find

$$\frac{d\mathbf{y}}{dt} = \mathbf{D} \cdot \mathbf{h}, \quad \text{i.e.,} \quad \frac{d\mathbf{h}}{dt} = \mathbf{D} \cdot \mathbf{h}.$$

This vector equation is equivalent to three linear differential equations that separate in the basis $\tilde{\mathbf{e}}_1, \tilde{\mathbf{e}}_2, \tilde{\mathbf{e}}_3$:

$$\frac{d\tilde{h}_i}{dt} = d_i \tilde{h}_i, \quad i = 1, 2, 3.$$

The rate of change of a unit length along the $\tilde{\mathbf{e}}_i$ axis at $t = 0$ is thus d_i. The vector field $\mathbf{D} \cdot \mathbf{h}$ is thus merely expanding or contracting along each of the axes $\tilde{\mathbf{e}}_i$—hence, the name "deformation." The rate of change of the volume of a box with sides of length $\tilde{h}_1, \tilde{h}_2, \tilde{h}_3$ parallel to the $\tilde{\mathbf{e}}_1, \tilde{\mathbf{e}}_2, \tilde{\mathbf{e}}_3$ axes is

$$\begin{aligned} \frac{d}{dt}(\tilde{h}_1 \tilde{h}_2 \tilde{h}_3) &= \left[\frac{d\tilde{h}_1}{dt}\right] \tilde{h}_2 \tilde{h}_3 + \tilde{h}_1 \left[\frac{d\tilde{h}_2}{dt}\right] \tilde{h}_3 + \tilde{h}_1 \tilde{h}_2 \left[\frac{d\tilde{h}_3}{dt}\right] \\ &= (d_1 + d_2 + d_3)(\tilde{h}_1 \tilde{h}_2 \tilde{h}_3). \end{aligned}$$

However, the trace of a matrix is invariant under orthogonal transformations. Hence,

$$d_1 + d_2 + d_3 = \text{trace of } \mathbf{D} = \text{trace of } \tfrac{1}{2}\left((\nabla \mathbf{u}) + (\nabla \mathbf{u})^T\right) = \operatorname{div} \mathbf{u}.$$

This confirms the fact proved in §**1.1** that volume elements change at a rate proportional to $\operatorname{div} \mathbf{u}$. Of course, the constant vector field $\mathbf{u}(\mathbf{x})$ in formula (1.2.1) induces a flow that is merely a translation by $\mathbf{u}(\mathbf{x})$. The other term, $\frac{1}{2}\boldsymbol{\xi}(\mathbf{x}) \times \mathbf{h}$, induces a flow

$$\frac{d\mathbf{h}}{dt} = \tfrac{1}{2}\boldsymbol{\xi}(\mathbf{x}) \times \mathbf{h}, \qquad (\mathbf{x} \text{ fixed}).$$

The solution of this linear differential equation is, by elementary vector calculus,

$$\mathbf{h}(t) = \mathbf{R}(t, \boldsymbol{\xi}(\mathbf{x}))\mathbf{h}(0),$$

where $\mathbf{R}(t, \boldsymbol{\xi}(\mathbf{x}))$ is the matrix that represents a rotation through an angle t about the axis $\boldsymbol{\xi}(\mathbf{x})$ (in the oriented sense). Because rigid motion leaves volumes invariant, the divergence of $\frac{1}{2}\boldsymbol{\xi}(\mathbf{x}) \times \mathbf{h}$ is zero, as may also be checked by noting that **S** has zero trace. This completes our derivation and discussion of the decomposition (1.2.1).

We remarked in §**1.1** that our assumptions so far have precluded any tangential forces, and thus any mechanism for starting or stopping rotation. Thus, intuitively, we might expect rotation to be conserved. Because rotation is intimately related to the vorticity as we have just shown, we can expect the vorticity to be involved. We shall now prove that this is so.

Let C be a simple closed contour in the fluid at $t = 0$. Let C_t be the contour carried along by the flow. In other words,

$$C_t = \varphi_t(C),$$

where φ_t is the fluid flow map discussed in §**1.1** (see Figure 1.2.1).

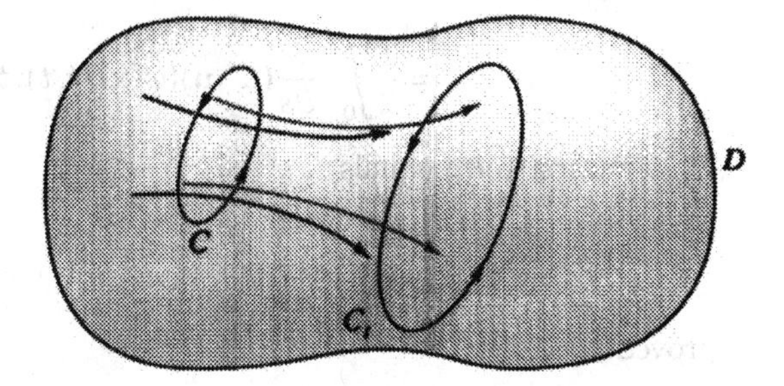

FIGURE 1.2.1. Kelvin's circulation theorem.

The ***circulation*** around C_t is defined to be the line integral

$$\Gamma_{C_t} = \oint_{C_t} \mathbf{u} \cdot ds.$$

Kelvin's Circulation Theorem *For isentropic flow without external forces, the circulation, Γ_{C_t} is constant in time.*

For example, we note that if the fluid moves in such a way that C_t shrinks in size, then the "angular" velocity around C_t increases. The proof of Kelvin's circulation theorem is based on a version of the transport theorem for curves.

Lemma *Let u be the velocity field of a flow and C a closed loop, with $C_t = \varphi_t(C)$ the loop transported by the flow* (Figure 1.2.1). *Then*

$$\frac{d}{dt}\int_{C_t} \mathbf{u} \cdot ds = \int_{C_t} \frac{D\mathbf{u}}{Dt}\, ds. \tag{1.2.5}$$

Proof Let $\mathbf{x}(s)$ be a parametrization of the loop C, $0 \le s \le 1$. Then a parameterization of C_t is $\varphi(\mathbf{x}(s), t), 0 \le s \le 1$. Thus, by definition of the line integral and the material derivative,

$$\begin{aligned}\frac{d}{dt}\int_{C_t} \mathbf{u}\cdot d\mathbf{s} &= \frac{d}{dt}\int_0^1 \mathbf{u}(\varphi(\mathbf{x}(s),t),t)\cdot\frac{\partial}{\partial s}\varphi(\mathbf{x}(s),t)\,ds\\ &= \int_0^1 \frac{D\mathbf{u}}{Dt}(\varphi(\mathbf{x}(s),t),t)\cdot\frac{\partial}{\partial s}\varphi(\mathbf{x}(s),t)\,ds\\ &\quad+\int_0^1 \mathbf{u}(\varphi(\mathbf{x}(s),t),t)\cdot\frac{\partial}{\partial t}\frac{\partial}{\partial s}\varphi(\mathbf{x}(s),t)\,ds.\end{aligned}$$

Because $\partial\varphi/\partial t = \mathbf{u}$, the second term equals

$$\begin{aligned}&\int_0^1 \mathbf{u}(\varphi(\mathbf{x}(s),t),t)\cdot\frac{\partial}{\partial s}\mathbf{u}(\varphi(\mathbf{x}(s),t),t)\,ds\\ &\qquad= \frac{1}{2}\int_0^1 \frac{\partial}{\partial s}(\mathbf{u}\cdot\mathbf{u})(\varphi(\mathbf{x}(s),t),t)\,ds = 0\end{aligned}$$

(since C_t is closed). The first term equals

$$\int_{C_t} \frac{D\mathbf{u}}{Dt}\,d\mathbf{s},$$

so the lemma is proved. ■

Proof of the Circulation Theorem Using the lemma and the fact that $D\mathbf{u}/Dt = -\nabla w$ (the flow is isentropic and without external forces), we find

$$\begin{aligned}\frac{d}{dt}\Gamma_{C_t} &= \frac{d}{dt}\int_{C_t}\mathbf{u}\,d\mathbf{s} = \int_{C_t}\frac{D\mathbf{u}}{Dt}\,d\mathbf{s}\\ &= -\int_{C_t}\nabla w\cdot d\mathbf{s} = 0 \quad \text{(since } C_t \text{ is closed).}\end{aligned}$$ ■

We now use Stokes' theorem, which will bring in the vorticity. If Σ is a surface whose boundary is an oriented closed oriented contour C, then Stokes' theorem yields (see Figure 1.2.2)

$$\Gamma_C = \int_C \mathbf{u}\cdot d\mathbf{s} = \iint_\Sigma (\nabla\times\mathbf{u})\cdot\mathbf{n}\,dA = \iint_\Sigma \boldsymbol{\xi}\cdot d\mathbf{A}.$$

Thus, as a corollary of the circulation theorem, we can conclude that *the flux of vorticity across a surface moving with the fluid is constant in time.*

By definition, a ***vortex sheet*** (or ***vortex line***) is a surface S (or a curve $\mathcal{L}$) that is tangent to the vorticity vector $\boldsymbol{\xi}$ at each of its points (Figure 1.2.3).

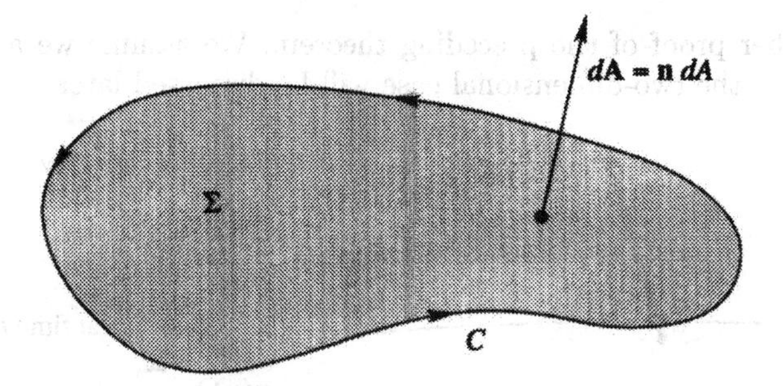

FIGURE 1.2.2. The circulation around C is the integral of the vorticity over Σ.

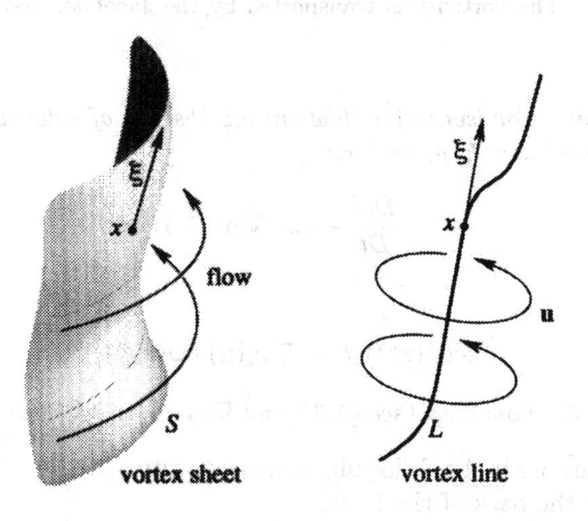

FIGURE 1.2.3. Vortex sheets and lines remain so under the flow.

Proposition *If a surface (or curve) moves with the flow of an isentropic fluid and is a vortex sheet (or line) at $t = 0$, then it remains so for all time.*

Proof Let $\mathbf{n}$ be the unit normal to S, so that at $t = 0$, $\boldsymbol{\xi} \cdot \mathbf{n} = 0$ by hypothesis. By the circulation theorem, the flux of $\boldsymbol{\xi}$ across any portion $\tilde{S} \subset S$ at a later time is also zero, *i.e.*,

$$\iint_{\tilde{S}_t} \boldsymbol{\xi} \cdot \mathbf{n}\, dA = 0.$$

It follows that $\boldsymbol{\xi} \cdot \mathbf{n} = 0$ identically on S_t, so S remains a vortex sheet.

One can show (using the implicit function theorem) that if $\boldsymbol{\xi}(\mathbf{x}) \neq \mathbf{0}$, then, locally, a vortex line is the intersection of two vortex sheets. ■

Next, we show that the vorticity (per unit mass), that is, $\boldsymbol{\omega} = \boldsymbol{\xi}/\rho$, is *propagated by the flow* (see Figure 1.2.4). This fact can also be used to give another proof of the preceding theorem. We assume we are in three dimensions; the two-dimensional case will be discussed later.

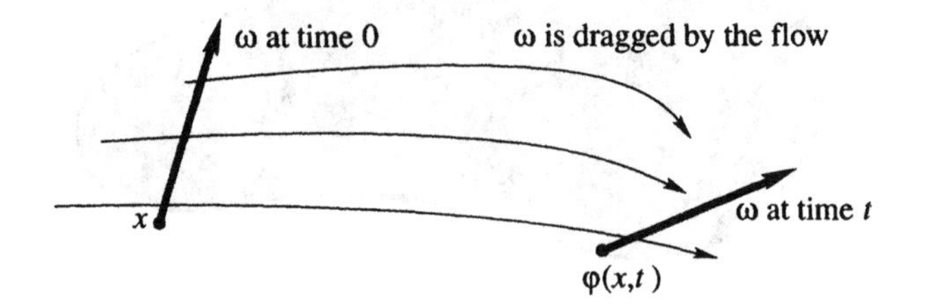

FIGURE 1.2.4. The vorticity is transported by the Jacobian matrix of the flow map.

Proposition *For isentropic flow (in the absence of external forces) with* $\boldsymbol{\xi} = \nabla \times \mathbf{u}$ *and* $\boldsymbol{\omega} = \boldsymbol{\xi}/\rho$, *we have*

$$\frac{D\boldsymbol{\omega}}{Dt} - (\boldsymbol{\omega} \cdot \nabla)\mathbf{u} = 0 \tag{1.2.6}$$

and

$$\boldsymbol{\omega}(\varphi(\mathbf{x}, t), t) = \nabla\varphi_t(\mathbf{x}) \cdot \boldsymbol{\omega}(\mathbf{x}, 0), \tag{1.2.7}$$

where φ_t *is the flow map (see* **§1.1***) and* $\nabla\varphi_t$ *is its Jacobian matrix.*

Proof Start with the following vector identity (see the table of vector identities at the back of the book)

$$\tfrac{1}{2}\nabla(\mathbf{u} \cdot \mathbf{u}) = \mathbf{u} \times \operatorname{curl} \mathbf{u} + (\mathbf{u} \cdot \nabla)\mathbf{u}.$$

Substituting this into the equations of motion yields

$$\frac{\partial \mathbf{u}}{\partial t} + \tfrac{1}{2}\nabla(\mathbf{u} \cdot \mathbf{u}) - \mathbf{u} \times \operatorname{curl} \mathbf{u} = -\nabla w.$$

Taking the curl and using the identity $\nabla \times \nabla f = 0$ gives

$$\frac{\partial \boldsymbol{\xi}}{\partial t} - \operatorname{curl}(\mathbf{u} \times \boldsymbol{\xi}) = \mathbf{0}.$$

Using the identity (also from the back of the book)

$$\operatorname{curl}(\mathbf{F} \times \mathbf{G}) = \mathbf{F} \operatorname{div} \mathbf{G} - \mathbf{G} \operatorname{div} \mathbf{F} + (\mathbf{G} \cdot \nabla)\mathbf{F} - (\mathbf{F} \cdot \nabla)\mathbf{G}$$

for the curl of a vector product, gives

$$\frac{\partial \boldsymbol{\xi}}{\partial t} - [(\mathbf{u}(\nabla \cdot \boldsymbol{\xi}) - \boldsymbol{\xi}(\nabla \cdot \mathbf{u}) + \boldsymbol{\xi} \cdot \nabla)\mathbf{u} - (\mathbf{u} \cdot \nabla)\boldsymbol{\xi}] = \mathbf{0},$$

that is,

$$\frac{D\boldsymbol{\xi}}{Dt} - (\boldsymbol{\xi} \cdot \nabla)\mathbf{u} + \boldsymbol{\xi}(\nabla \cdot \mathbf{u}) = \mathbf{0}, \tag{1.2.8}$$

since $\boldsymbol{\xi}$ is divergence free. Also,

$$\frac{D\boldsymbol{\omega}}{Dt} = \frac{D}{Dt}\left(\frac{\boldsymbol{\xi}}{\rho}\right) = \frac{1}{\rho}\frac{D\boldsymbol{\xi}}{Dt} + \frac{\boldsymbol{\xi}}{\rho}(\nabla \cdot \mathbf{u}) \tag{1.2.9}$$

by the continuity equation. Substitution of (1.2.8) into (1.2.9) yields (1.2.6).
To prove (1.2.7), let

$$\mathbf{F}(\mathbf{x},t) = \boldsymbol{\omega}(\varphi(\mathbf{x},t),t) \quad \text{and} \quad \mathbf{G}(\mathbf{x},t) = \nabla\varphi_t(\mathbf{x}) \cdot \boldsymbol{\omega}(\mathbf{x},0).$$

By (1.2.6), $\partial\mathbf{F}/\partial t = (\mathbf{F} \cdot \nabla)\mathbf{u}$. On the other hand, by the chain rule:

$$\begin{aligned}\frac{\partial \mathbf{G}}{\partial t} &= \nabla\left[\frac{\partial\varphi}{\partial t}(\mathbf{x},t)\right] \cdot \boldsymbol{\omega}(\mathbf{x},0) = \nabla(\mathbf{u}(\varphi(\mathbf{x},t),t)) \cdot \boldsymbol{\omega}(\mathbf{x},0) \\ &= (\nabla\mathbf{u}) \cdot \nabla\varphi_t(\mathbf{x}) \cdot \boldsymbol{\omega}(\mathbf{x},0) = (\mathbf{G} \cdot \nabla)\mathbf{u}\end{aligned}$$

Thus, $\mathbf{F}$ and $\mathbf{G}$ satisfy the same linear first-order differential equation. Because they coincide at $t = 0$ and solutions are unique, they are equal. ■

The reader may wish to compare (1.2.7) with the formula

$$\rho(\mathbf{x},0) = \rho(\varphi(\mathbf{x},t),t)J(\mathbf{x},t) \tag{1.2.10}$$

proved in §1.1.
As an exercise, we invite the reader to prove the preservation of vortex sheets and lines by the flow using (1.2.7) and (1.2.10).

For two-dimensional flow, where $\mathbf{u} = (u, v, 0)$, $\boldsymbol{\xi}$ has only one component; $\boldsymbol{\xi} = (0, 0, \xi)$. The circulation theorem now states that if Σ_t is any region in the plane that is moving with the fluid, then

$$\int_{\Sigma_t} \xi \, dA = \text{constant in time.} \tag{1.2.11}$$

In fact, one can say more using (1.2.7). In two dimensions, (1.2.7) specializes to

$$\frac{\xi}{\rho}(\varphi(\mathbf{x},t),t) = \frac{\xi}{\rho}(\mathbf{x},0), \tag{1.2.7$'$}$$

that is, ξ/ρ is propagated as a scalar by the flow. Employing (1.2.10) and the change of variables theorem gives (1.2.11) as a special case.

In three-dimensional flows, the relation (1.2.7) allows rather complicated behavior. We shall now discuss the three-dimensional geometry a bit further.

A ***vortex tube*** consists of a two-dimensional surface S that is nowhere tangent to $\boldsymbol{\xi}$, with vortex lines drawn through each point of the bounding curve C of S. These vortex lines are integral curves of $\boldsymbol{\xi}$ and are extended as far as possible in each direction. See Figure 1.2.5.

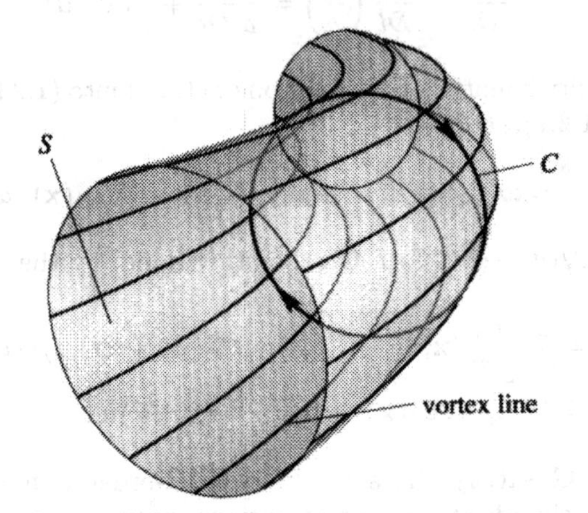

FIGURE 1.2.5. A vortex tube consists of vortex lines drawn through points of C.

In fluid mechanics it is customary to be sloppy about this definition and make tacit assumptions to the effect that the tube really "looks like" a tube. More precisely, we assume S is diffeomorphic to a disc (i.e., related to a disc by a one-to-one invertible differentiable transformation) and that the resulting tube is diffeomorphic to the product of the disc and the real line. This tacitly assumes that $\boldsymbol{\xi}$ has no zeros (of course, $\boldsymbol{\xi}$ *could* have zeros!).

Helmholtz's Theorem *Assume the fluid is isentropic. Then*

(*a*) *If C_1 and C_2 are any two curves encircling the vortex tube, then*

$$\int_{C_1} \mathbf{u} \cdot d\mathbf{s} = \int_{C_2} \mathbf{u} \cdot d\mathbf{s}.$$

This common value is called the ***strength*** *of the vortex tube.*

(*b*) *The strength of the vortex tube is constant in time, as the tube moves with the fluid.*

Proof (a) Let C_1 and C_2 be oriented as in Figure 1.2.6.

The lateral surface of the vortex tube enclosed between C_1 and C_2 is denoted by S, and the end faces with boundaries C_1 and C_2 are denoted by S_1 and S_2, respectively. Since $\boldsymbol{\xi}$ is tangent to the lateral surface, S is a

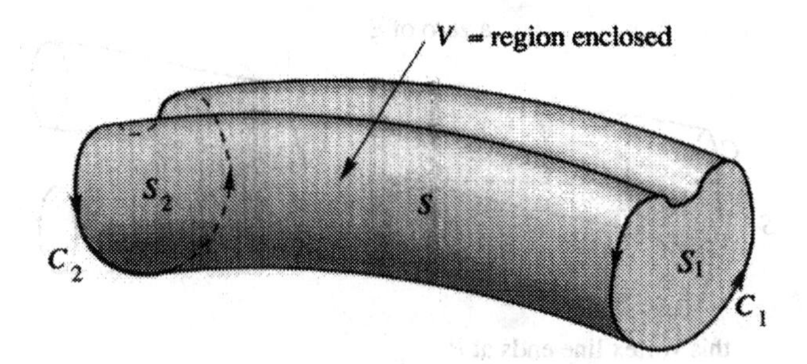

FIGURE 1.2.6. A vortex tube enclosed between two curves, C_1 and C_2.

vortex sheet. Let V denote the region of the vortex tube between C_1 and C_2 and $\Sigma = S \cup S_1 \cup S_2$ denote the boundary of V. By Gauss' theorem,

$$0 = \int_V \nabla \cdot \boldsymbol{\xi}\, dx = \int_\Sigma \boldsymbol{\xi} \cdot d\mathbf{A} = \int_{S_1 \cup S_2} \boldsymbol{\xi} \cdot d\mathbf{A} + \int_S \boldsymbol{\xi} \cdot d\mathbf{A}.$$

By Stokes' theorem

$$\int_{C_1} \mathbf{u} \cdot d\mathbf{s} = \int_{S_1} \boldsymbol{\xi} \cdot d\mathbf{A} \quad \text{and} \quad \int_{C_2} \mathbf{u} \cdot d\mathbf{s} = -\int_{S_2} \boldsymbol{\xi} \cdot d\mathbf{A},$$

so (a) holds. Part (b) now follows from Kelvin's circulation theorem. ■

Observe that if a vortex tube gets stretched and its cross-sectional area decreases, then the magnitude of $\boldsymbol{\xi}$ must increase. Thus, the stretching of vortex tubes can increase vorticity, but it cannot create it.

A vortex tube with nonzero strength cannot "end" in the interior of the fluid. It either forms a ring (such as the smoke from a cigarette), extends to infinity, or is attached to a solid boundary. The usual argument supporting this statement goes like this: suppose the tube ended at a certain cross section S, inside the fluid. Because the tube cannot be extended, we must have $\boldsymbol{\xi} = \mathbf{0}$ on C_1. Thus, the strength is zero—a contradiction.

This "proof" is hopelessly incomplete. First of all, why should a vortex tube end in a nice regular way on a surface? Why can't it split in two, as in Figure 1.2.7? There is no *a priori* reason why this sort of thing cannot happen unless we merely exclude it by tacit assumption.[4]. In particular, note that the assertion often made that a vortex line cannot end in the fluid is clearly false if we allow $\boldsymbol{\xi}$ to have zeros and probably is false even if $\boldsymbol{\xi}$ has no zeros (an orbit of a vector field can wander around forever without accumulating at an endpoint—as with a line with irrational slope on a torus)

[4]H. Lamb [1895] *Mathematical Theory of the Motion of Fluids,* Cambridge Univ. Press, p. 149.

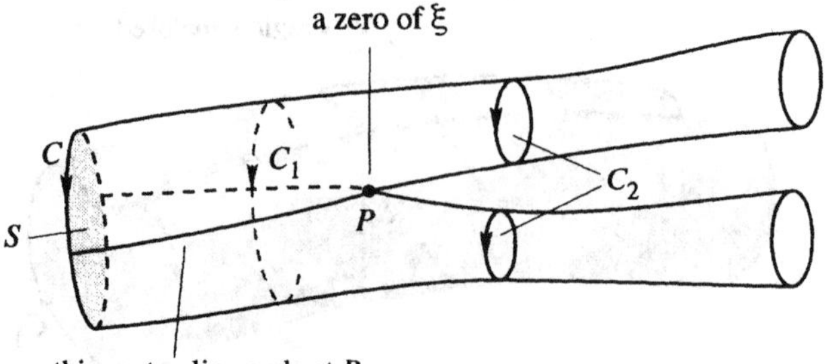

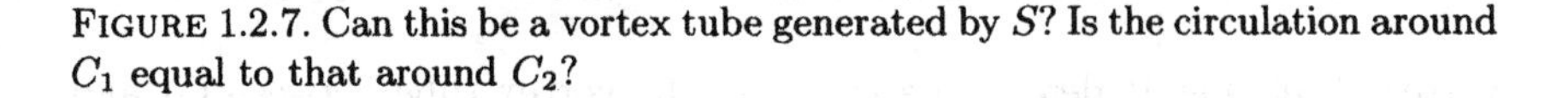
FIGURE 1.2.7. Can this be a vortex tube generated by S? Is the circulation around C_1 equal to that around C_2?

Thus, our assertion about vortex tubes "ending" is correct if we interpret "ending" properly. But the reader is cautioned that this may not be all that can happen, and that this time-honored statement is not at all a proved theorem.

The difference between the two-dimensional and three-dimensional conservation laws for vorticity is very important. The conservation of vorticity (1.2.7)′ in two dimensions is a helpful tool in establishing a rigorous theory of existence and uniqueness of the Euler (and later Navier–Stokes) equations. The lack of the same kind of conservation in three dimensions is a major obstacle to the rigorous understanding of crucial properties of the solutions of the equations of fluid dynamics. The main point here is to get existence theorems for *all time*. At the moment, it is known only in two dimension that all time smooth solutions exist.

Our last main goal in this section is to develop the vorticity equation somewhat further for the important special case of incompressible flow. For two-dimensional homogeneous incompressible flow, the ***vorticity equation*** is

$$\frac{D\xi}{Dt} = \partial_t \xi + (\mathbf{u} \cdot \nabla)\xi = 0, \tag{1.2.12}$$

where $\xi = \xi(x, y, t) = \partial_x v - \partial_y u$ is the (scalar) vorticity field of the flow and u, v are the components of $\mathbf{u}$. Assume that the flow is contained in some plane domain D with a fixed boundary ∂D, with the boundary condition

$$\mathbf{u} \cdot \mathbf{n} = 0 \qquad \text{on } \partial D, \tag{1.2.13}$$

where $\mathbf{n}$ is the unit outward normal to ∂D. Let us assume D is simply connected (i.e., has no "holes"). Then, by incompressibility, $\partial_x u = -\partial_y v$, and so from vector calculus there is a scalar function $\psi(x, y, t)$ on D unique up to an additive constant such that

$$u = \partial_y \psi \quad \text{and} \quad v = -\partial_x \psi. \tag{1.2.14}$$

The function ψ is the ***stream function*** for fixed t; streamlines lie on level curves of ψ. Indeed, let $(x(s), y(s))$ be a streamline, so $x' = u(x,y)$ and $y' = v(x,y)$. Then

$$\frac{d}{ds}\psi(x(s), y(s), t) = \partial_x\psi \cdot x' + \partial_y\psi \cdot y' = -vu + uv = 0.$$

In particular, by (1.2.13), ∂D lies on a level curve of ψ, and we can adjust the constant so that

$$\psi(x,y,t) = 0 \qquad \text{for } (x,y) \in \partial D.$$

This convention and (1.2.14) determine ψ uniquely. (∂D need not be a whole streamline, but can be composed of streamlines separated by zeros of $\mathbf{u}$, that is, by ***stagnation points***.) The scalar vorticity is now given by

$$\xi = \partial_x v - \partial_y u = -\partial_x^2\psi - \partial_y^2\psi = -\Delta\psi,$$

where $\Delta = \partial_x^2 + \partial_y^2$ is the Laplace operator in the plane.

We can summarize the equations for ξ for two-dimensional incompressible flow as follows:

$$\left.\begin{array}{ll} & \frac{D\xi}{Dt} \equiv \partial_t\xi + (\mathbf{u}\cdot\nabla)\xi = 0, \\ & \Delta\psi = -\xi, \\ \text{with} & \\ & \psi = 0 \quad \text{on } \partial D, \\ \text{and with} & \\ & u = \partial_y\psi \quad \text{and} \quad v = -\partial_x\psi. \end{array}\right\} \qquad (1.2.15)$$

These equations completely determine the flow. Note that given ξ, the function ψ is determined by $\Delta\psi = -\xi$ and the boundary conditions, and hence $\mathbf{u}$ by the last equations in (1.2.15). Thus, ξ completely determines $\partial_t\xi$ and hence the evolution of ξ and, through it, ψ and $\mathbf{u}$.

Another remark is useful:

$$\begin{aligned}(\mathbf{u}\cdot\nabla)\xi &= u\partial_x\xi + v\partial_y\xi = (\partial_y\psi)(\partial_x\xi) - (\partial_x\psi)(\partial_y\xi) \\ &= \det\begin{bmatrix} \partial_x\xi & \partial_y\xi \\ \partial_x\psi & \partial_y\psi \end{bmatrix} = J(\xi,\psi),\end{aligned}$$

the Jacobian of ξ and ψ. Thus, *the flow is stationary (time independent) if and only if ξ and ψ are functionally dependent.* (If functional dependence holds at one instant it will hold for all time as a consequence.)

Example Suppose at $t = 0$ the stream function $\psi(x,y)$ is a function only of the radial distance $r = (x^2 + y^2)^{1/2}$. Thus, the streamlines are concentric circles. Write $\psi(x,y) = \psi(r)$ and assume $\psi_r > 0$. The velocity vector is given by

$$u = \partial_y \psi = \partial_r \psi \partial_y r = \frac{y}{r} \partial_r \psi, \tag{1.2.16}$$

$$v = -\partial_x \psi = -\partial_r \psi \partial_x r = -\frac{x}{r} \partial_r \psi, \tag{1.2.17}$$

that is, $\mathbf{u}$ is tangent to the circle of radius r with magnitude $|\partial_r \psi|$ and oriented clockwise if $\psi_r > 0$ and counterclockwise if $\psi_r < 0$. Next, observe that

$$\xi = -\Delta\psi = -\frac{1}{r}\frac{\partial}{\partial r}\left(r\frac{\partial \psi}{\partial r}\right),$$

a function of r alone. Because $\psi_r \neq 0, r$ is a function of ψ so ξ is also a function of ψ. Thus, $J(\xi, \psi) = 0$. Hence, motion in concentric circles with $\mathbf{u}$ defined as above is a solution of the two-dimensional *stationary* incompressible equations of ideal flow.

For three-dimensional incompressible ideal flow, the analogue of (1.2.15) is

$$\left.\begin{aligned} &\frac{D\boldsymbol{\xi}}{Dt} - (\boldsymbol{\xi}\cdot\nabla)\mathbf{u} = 0, \\ &\Delta\mathbf{A} = -\boldsymbol{\xi}, \qquad \operatorname{div}\mathbf{A} = 0, \\ &\mathbf{u} = \nabla\times\mathbf{A}. \end{aligned}\right\} \tag{1.2.18}$$

Here we used $\nabla\cdot\mathbf{u} = 0$ to write $\mathbf{u} = \nabla\times\mathbf{A}$, where $\operatorname{div}\mathbf{A} = 0$. (This requires not that D be simply connected, but that it not have any "solid holes" in it; for instance, if D is convex, this will hold.) Then

$$\boldsymbol{\xi} = \operatorname{curl}\mathbf{u} = \operatorname{curl}(\operatorname{curl}\mathbf{A}) = -\Delta A + \nabla(\operatorname{div}\mathbf{A}) = -\Delta\mathbf{A}.$$

One of the troubles with (1.2.18) is that given $\boldsymbol{\xi}$, the vector field $\mathbf{A}$ is not uniquely determined (we cannot impose boundary condition such as $\mathbf{A} = 0$ on ∂D because $\mathbf{A}$ need not be constant on ∂D as was the case with ψ). ♦

Exercises

◇ **Exercise 1.2-1** Derive a formula akin to the transport theorem and Kelvin's circulation theorem for

$$\frac{d}{dt}\int_{S_t} \mathbf{v}\cdot\mathbf{n}\,dA,$$

where S_t is a *moving surface* and $\mathbf{v}$ is a vector field.

◇ **Exercise 1.2-2** ***Couette flow.*** Let Ω be the region between two concentric cylinders of radii R_1 and R_2, where $R_1 < R_2$. Define $\mathbf{v}$ in cylindrical coordinates by

$$v_r = 0, \qquad v_z = 0,$$

and

$$v_\theta = \frac{A}{r} + Br,$$

where

$$A = -\frac{R_1^2 R_2^2(\omega_2 - \omega_1)}{R_2^2 - R_2^1} \quad \text{and} \quad B = -\frac{R_1^2\omega_1 - R_2^2\omega_2}{R_2^2 - R_1^2}.$$

Show that

(i) $\mathbf{v}$ is a stationary solution of Euler's equations with $\rho = 1$;

(ii) $\boldsymbol{\omega} = \nabla \times \mathbf{v} = (0, 0, 2B)$;

(iii) the deformation tensor is

$$D = -\frac{A}{r^2}\begin{bmatrix} 0 & 1 \\ 1 & 0 \end{bmatrix}$$

and discuss its physical meaning;

(iv) the angular velocity of the flow on the two cylinders is ω_1 and ω_2.

1.3 The Navier–Stokes Equations

In §**1.1** we defined an ideal fluid as one in which forces across a surface were normal to that surface. We now consider more general fluids. To understand the need for the generalization beyond the examples already given, consider the situation shown in Figure 1.3.1. Here the velocity field $\mathbf{u}$ is parallel to a surface S but jumps in magnitude either suddenly or rapidly as we cross S. If the forces are all normal to S, there will be no transfer of momentum between the fluid volumes denoted by B and B' in Figure 1.3.1. However, if we remember the kinetic theory of matter, we see that this is actually unreasonable. Faster molecules from above S will diffuse across S and impart momentum to the fluid below, and, likewise, slower molecules from below S will diffuse across S to slow down the fluid above S. For reasonably fast changes in velocity over short distance, this effect is important.[5]

We thus change our previous definition. Instead of assuming that

$$\text{force on } S \text{ per unit area} = -p(\mathbf{x}, t)\mathbf{n},$$

[5] For more information, see J. Jeans [1867] *An Introduction to the Kinetic Theory of Gases*, Cambridge Univ. Press.

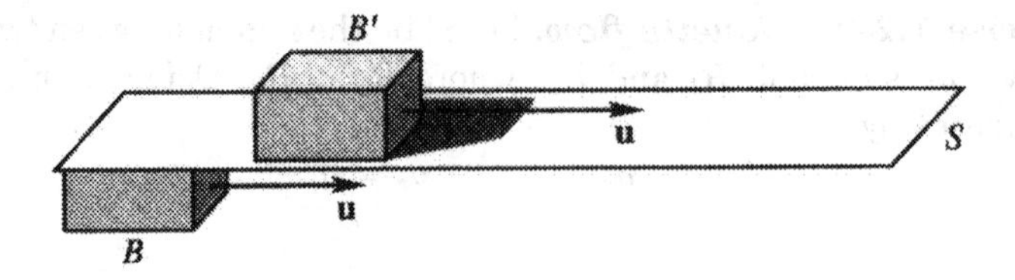

FIGURE 1.3.1. Faster molecules in B' can diffuse across S and impart momentum to B.

where $\mathbf{n}$ is the normal to S, we now assume that

$$\text{force on } S \text{ per unit area} = -p(\mathbf{x}, t)\mathbf{n} + \boldsymbol{\sigma}(\mathbf{x}, t) \cdot \mathbf{n}, \tag{1.3.1}$$

where $\boldsymbol{\sigma}$ is a *matrix* called the ***stress tensor***, about which some assumptions will have to be made. The new feature is that $\boldsymbol{\sigma} \cdot \mathbf{n}$ need not be parallel to $\mathbf{n}$. The separation of the forces into pressure and other forces in (1.3.1) is somewhat ambiguous because $\boldsymbol{\sigma} \cdot \mathbf{n}$ may contain a component parallel to $\mathbf{n}$. This issue will be resolved later when we give a more definite functional form to $\boldsymbol{\sigma}$.

As before, Newton's second law states that the rate of change of any moving portion of fluid W_t equals the force acting on it (balance of momentum):

$$\frac{d}{dt} \int_{W_t} \rho \mathbf{u} \, dV = - \int_{\partial W_t} (p \cdot \mathbf{n} - \boldsymbol{\sigma} \cdot \mathbf{n}) \, dA$$

(compare (BM3) in §**1.1**). Thus, we see that $\boldsymbol{\sigma}$ modifies the transport of momentum across the boundary of W_t. We will choose $\boldsymbol{\sigma}$ so that it approximates in a reasonable way the transport of momentum by molecular motion.

One can legitimately ask why the force (1.3.1) acting on S should be a *linear* function of $\mathbf{n}$. In fact, if one just assumes the force is a continuous function of $\mathbf{n}$, then, using balance of momentum, one can *prove* it is linear in $\mathbf{n}$. This result is called ***Cauchy's Theorem***.[6]

Our assumptions on $\boldsymbol{\sigma}$ are as follows:

1. $\boldsymbol{\sigma}$ *depends linearly on the velocity gradients* $\nabla \mathbf{u}$ that is, $\boldsymbol{\sigma}$ is related to $\nabla \mathbf{u}$ by some linear transformation at each point.

2. $\boldsymbol{\sigma}$ *is invariant under rigid body rotations*, that is, if $\mathbf{U}$ is an orthogonal matrix,

$$\boldsymbol{\sigma}(\mathbf{U} \cdot \nabla \mathbf{u} \cdot \mathbf{U}^{-1}) = \mathbf{U} \cdot \boldsymbol{\sigma}(\nabla \mathbf{u}) \cdot \mathbf{U}^{-1}.$$

[6]For a proof and further references, see, for example, Marsden and Hughes [1994].

This is reasonable, because when a fluid undergoes a rigid body rotation, there should be no diffusion of momentum.

3. *$\boldsymbol{\sigma}$ is symmetric*. This property can be deduced as a consequence of balance of angular momentum.[7]

Since $\boldsymbol{\sigma}$ is symmetric, if follows from properties 1 and 2 that $\boldsymbol{\sigma}$ can depend only on the symmetric part of $\nabla \mathbf{u}$; that is, on the deformation $\mathbf{D}$. Because $\boldsymbol{\sigma}$ is a linear function of $\mathbf{D}$, $\boldsymbol{\sigma}$ and $\mathbf{D}$ commute and so can be simultaneously diagonalized. Thus, the eigenvalues of $\boldsymbol{\sigma}$ are linear functions of those of $\mathbf{D}$. By property 2, they must also be symmetric because we can choose $\mathbf{U}$ to permute two eigenvalues of $\mathbf{D}$ (by rotating through an angle $\pi/2$ about an eigenvector), and this must permute the corresponding eigenvalues of $\boldsymbol{\sigma}$. The only linear functions that are symmetric in this sense are of the form

$$\sigma_i = \lambda(d_1 + d_2 + d_3) + 2\mu d_i, \qquad i = 1, 2, 3,$$

where σ_i are the eigenvalues of $\boldsymbol{\sigma}$, and d_i are those of $\mathbf{D}$. This defines the constants λ and μ. Recalling that $d_1 + d_2 + d_3 = \operatorname{div} \mathbf{u}$, we can use property 2 to transform σ_i back to the usual basis and deduce that

$$\boldsymbol{\sigma} = \lambda(\operatorname{div} \mathbf{u})\mathbf{I} + 2\mu \mathbf{D}, \tag{1.3.2}$$

where $\mathbf{I}$ is the identity. We can rewrite this by putting all the trace in one term:

$$\boldsymbol{\sigma} = 2\mu[\mathbf{D} - \tfrac{1}{3}(\operatorname{div} \mathbf{u})\mathbf{I}] + \zeta(\operatorname{div} \mathbf{u})\mathbf{I} \tag{1.3.2$'$}$$

where μ is the ***first coefficient of viscosity***, and $\zeta = \lambda + \frac{2}{3}\mu$ is the ***second coefficient of viscosity.***

If we employ the transport theorem and the divergence theorem, as we did in connection with (BM3), balance of momentum yields the ***Navier–Stokes equations***,

$$\boxed{\rho \frac{D\mathbf{u}}{Dt} = -\nabla p + (\lambda + \mu)\nabla(\operatorname{div} \mathbf{u}) + \mu \Delta \mathbf{u}} \tag{1.3.3}$$

where

$$\Delta \mathbf{u} = \left(\frac{\partial^2}{\partial x^2} + \frac{\partial^2}{\partial y^2} + \frac{\partial^2}{\partial z^2} \right) \mathbf{u}$$

is the Laplacian of $\mathbf{u}$. Together with the equation of continuity and an energy equation, (1.3.3) completely describes the flow of a compressible viscous fluid.

[7] Op. cit.

In the case of incompressible homogeneous flow $\rho = \rho_0 =$ constant, the complete set of equations becomes the ***Navier–Stokes equations for incompressible flow***,

$$\boxed{\begin{aligned} \frac{D\mathbf{u}}{Dt} &= -\operatorname{grad} p' + \nu\Delta\mathbf{u} \\ \operatorname{div}\mathbf{u} &= 0 \end{aligned}} \tag{1.3.4}$$

where $\nu = \mu/\rho_0$ is the coefficient of *kinematic viscosity*, and $p' = p/\rho_0$.

These equations are supplemented by boundary conditions. For Euler's equations for ideal flow we use $\mathbf{u} \cdot \mathbf{n} = 0$, that is, fluid does not cross the boundary but may move tangentially to the boundary. For the Navier–Stokes equations, the extra term $\nu\Delta\mathbf{u}$ raises the number of derivatives of $\mathbf{u}$ involved from one to two. For both experimental and mathematical reasons, this is accompanied by an increase in the number of boundary conditions. For instance, on a solid wall at rest we add the condition that the tangential velocity also be zero (the "no-slip condition"), so the full boundary conditions are simply

$$\mathbf{u} = \mathbf{0} \text{ on solid walls at rest.}$$

The mathematical need for extra boundary conditions hinges on their role in proving that the equations are well posed; that is, that a unique solution exists and depends continuously on the initial data. In three dimensions, it is known that smooth solutions to the incompressible equations exist for a short time and depend continuously on the initial data.[8] It is a major open problem in fluid mechanics to prove or disprove that solutions of the incompressible equations exist for all time. In two dimensions, solutions are known to exist for all time, for both viscous and inviscid flow[9]. In any case, adding the tangential boundary condition is crucial for viscous flow.

The physical need for the extra boundary conditions comes from simple experiments involving flow past a solid wall. For example, if dye is injected into flow down a pipe and is carefully watched near the boundary, one sees that the velocity approaches zero at the boundary to a high degree of precision. The no-slip condition is also reasonable if one contemplates the physical mechanism responsible for the viscous terms, namely,

[8] For a review of much of what is known, see O. A. Ladyzhenskaya [1969] *The Mathematical Theory of Viscous Incompressible Flow*, Gordon and Breach. See also R. Temam [1977] *Navier–Stokes Equations*, North Holland.

[9] Op. cit. and W. Wolibner, *Math. Zeit.* **37** [1933], 698–726; V. Judovich, *Mat. Sb. N.S.* **64** [1964], 562–588; and T. Kato, *Arch. Rational Mech. Anal.* **25** [1967], 188–200.

molecular diffusion. Our opening example indicates that molecular interaction between the solid wall with zero tangential velocity (or zero average velocity on the molecular level) should impart the same condition to the immediately adjacent fluid.

Another crucial feature of the boundary condition $\mathbf{u} = \mathbf{0}$ is that it provides a mechanism by which a boundary can produce vorticity in the fluid. We shall describe this in some detail in Chapter **2**.

Next, we shall discuss some scaling properties of the Navier–Stokes equations with the aim of introducing a parameter (the Reynolds number) that measures the effect of viscosity on the flow.

For a given problem, let L be a ***characteristic length*** and U a ***characteristic velocity***. These numbers are chosen in a somewhat arbitrary way. For example, if we consider flow past a sphere, L could be either the radius or the diameter of the sphere, and U could be the magnitude of the fluid velocity at infinity. L and U are merely reasonable length and velocity scales typical of the flow at hand. Their choice then determines a time scale by $T = L/U$.

We can measure $\mathbf{x}, \mathbf{u}$, and t as fractions of these scales, by changing variables and introducing the following dimensionless quantities

$$\mathbf{u}' = \frac{\mathbf{u}}{U}, \quad \mathbf{x}' = \frac{\mathbf{x}}{L}, \quad \text{and} \quad t' = \frac{t}{T}.$$

The x component of the (homogeneous) incompressible Navier–Stokes equation is

$$\frac{\partial u}{\partial t} + u\frac{\partial u}{\partial x} + v\frac{\partial u}{\partial y} + w\frac{\partial u}{\partial z} = -\frac{1}{\rho_0}\frac{\partial p}{\partial x} + \nu\left[\frac{\partial^2 u}{\partial x^2} + \frac{\partial^2 u}{\partial y^2} + \frac{\partial^2 u}{\partial z^2}\right].$$

The change of variables produces

$$\begin{aligned}\frac{\partial(u'U)}{\partial t'}\frac{\partial t'}{\partial t} &+ Uu'\frac{\partial(u'U)}{\partial x'}\frac{\partial x'}{\partial x} + Uv'\frac{\partial(u'U)}{\partial y'}\frac{\partial y'}{\partial y} + Uw'\frac{\partial(u'U)}{\partial z'}\frac{\partial w'}{\partial z} \\ &= -\frac{1}{\rho_0}\frac{\partial p}{\partial x'}\frac{\partial x'}{\partial x} + \nu\left[\frac{\partial^2(u'U)}{\partial(Lx')^2} + \frac{\partial^2(u'U)}{\partial(Ly')^2} + \frac{\partial^2(u'U)}{\partial(Lz')^2}\right],\end{aligned}$$

$$\begin{aligned}\left[\frac{U^2}{L}\right]&\left[\frac{\partial u'}{\partial t'} + u'\frac{\partial u'}{\partial x'} + v'\frac{\partial u'}{\partial y'} + w'\frac{\partial u'}{\partial z'}\right] \\ &= -\left[\frac{U^2}{L}\right]\frac{\partial(p/(\rho_0 U^2))}{\partial x'} + \left[\frac{U}{L^2}\right]\nu\left[\frac{\partial^2 u'}{\partial x'^2} + \frac{\partial^2 u'}{\partial y'^2} + \frac{\partial^2 u'}{\partial z'^2}\right].\end{aligned}$$

Similar equations hold for the y and z components. If we combine all three components and divide out by U^2/L, we obtain

$$\frac{\partial \mathbf{u}'}{\partial t'} + (\mathbf{u}' \cdot \nabla')\mathbf{u}' = -\operatorname{grad}\mathbf{p}' + \frac{\nu}{LU}\Delta'\mathbf{u}', \tag{1.3.5}$$

where $p' = p/(\rho_0 U^2)$. Incompressibility still reads

$$\operatorname{div} \mathbf{u}' = 0.$$

The equations (1.3.5) are the Navier–Stokes equations in dimensionless variables. We define the ***Reynolds number*** R to be the dimensionless number

$$R = \frac{LU}{\nu}.$$

For example, consider two flows past two spheres centered at the origin but with differing radii, one with a fluid where $U_\infty = 10$ km/hr past a sphere of radius 10 m and the other with the same fluid but with $U_\infty = 100$ km/hr and radius $= 1$ m. If we choose L to be the radius and U to be the velocity U_∞ at infinity, then the Reynolds number is the same for each flow. The equations satisfied by the dimensionless variables are thus identical for the two flows.

Two flows with the same geometry and the same Reynolds number are called ***similar***. More precisely, let $\mathbf{u}_1$ and $\mathbf{u}_2$ be two flows on regions D_1 and D_2 that are related by a scale factor λ so that $L_1 = \lambda L_2$. Let choices of U_1 and U_2 be made for each flow, and let the viscosities be ν_1 and ν_2 respectively. If

$$R_1 = R_2, \quad \text{i.e.,} \quad \frac{L_1 U_1}{\nu_1} = \frac{L_2 U_2}{\nu_2},$$

then the dimensionless velocity fields $\mathbf{u}_1'$ and $\mathbf{u}_2'$ satisfy exactly the same equation on the same region. Thus, we can conclude that $\mathbf{u}_1$ may be obtained from a suitably rescaled solution $\mathbf{u}_2$; in other words, $\mathbf{u}_1$ and $\mathbf{u}_2$ are similar.

This idea of the similarity of flows is used in the design of experimental models. For example, suppose we are contemplating a new design for an aircraft wing and we wish to know the behavior of a fluid flow around it. Rather than build the wing itself, it may be faster and more economical to perform the initial tests on a scaled-down version. We design our model so that it has the same geometry as the full-scale wing and we choose values for the undisturbed velocity, coefficient of viscosity, and so on, such that the Reynolds number for the flow in our experiment matches that of the actual flow. We can then expect the results of our experiment to be relevant to the actual flow over the full-scale wing.

We shall be especially interested in cases where R is large. We stress that one cannot say that if ν is small, then viscous effects are unimportant, because such a comment fails to consider the other dimensions of the problem, that is, "ν is small" *is not* a physically meaningful statement unless some scaling is chosen, but "$1/R$ is small" *is* a meaningful statement.

As with incompressible ideal flow, the pressure p in incompressible viscous flow is determined through the equation $\operatorname{div}\mathbf{u} = 0$. We now shall explore the role of the pressure in incompressible flow in more depth. Let D be a region in space (or in the plane) with smooth boundary ∂D.

We shall use the following decomposition theorem.

Helmholtz–Hodge Decomposition Theorem *A vector field* $\mathbf{w}$ *on* D *can be uniquely decomposed in the form*

$$\mathbf{w} = \mathbf{u} + \operatorname{grad} p, \tag{1.3.6}$$

where u has zero divergence and is parallel to ∂D*; that is,* $\mathbf{u}\cdot\mathbf{n} = 0$ *on* ∂D.

Proof First of all, we establish the orthogonality relation

$$\int_D \mathbf{u}\cdot\operatorname{grad} p\, dV = 0.$$

Indeed, by the identity

$$\operatorname{div}(p\mathbf{u}) = (\operatorname{div}\mathbf{u})p + \mathbf{u}\cdot\operatorname{grad} p,$$

the divergence theorem, and $\operatorname{div}\mathbf{u} = 0$, we get

$$\int_D \mathbf{u}\cdot\operatorname{grad} p\, dV = \int_D \operatorname{div}(p\mathbf{u})\, dV = \int_{\partial D} p\mathbf{u}\cdot\mathbf{n}\, dA = 0,$$

because $\mathbf{u}\cdot\mathbf{n} = 0$ on ∂D. We use this orthogonality to prove uniqueness. Suppose the $\mathbf{w} = \mathbf{u}_1 + \operatorname{grad} p_1 = \mathbf{u}_2 + \operatorname{grad} p_2$. Then

$$0 = \mathbf{u}_1 - \mathbf{u}_2 + \operatorname{grad}(p_1 - p_2).$$

Taking the inner product with $\mathbf{u}_1 - \mathbf{u}_2$ and integrating, we get

$$0 = \int_D \left\{\|\mathbf{u}_1 - \mathbf{u}_2\|^2 + (\mathbf{u}_1 - \mathbf{u}_2)\cdot\operatorname{grad}(p_1 - p_2)\right\} dV = \int_D \|\mathbf{u}_1 - \mathbf{u}_2\|^2\, dV$$

by the orthogonality relation. It follows that $\mathbf{u}_1 = \mathbf{u}_2$, and so, $\operatorname{grad} p_1 = \operatorname{grad} p_2$ (which is the same thing as $p_1 = p_2 + \text{constant}$).

If $\mathbf{w} = \mathbf{u} + \operatorname{grad} p$, notice that $\operatorname{div}\mathbf{w} = \operatorname{div}\operatorname{grad} p = \Delta p$ and that $\mathbf{w}\cdot\mathbf{n} = \mathbf{n}\cdot\operatorname{grad} p$. We use this remark to prove existence. Indeed, given $\mathbf{w}$, let p be defined by the solution to the Neumann problem

$$\Delta p = \operatorname{div}\mathbf{w} \quad \text{in } D, \quad \text{with} \quad \frac{\partial p}{\partial n} = \mathbf{w}\cdot\mathbf{n} \quad \text{on } \partial D.$$

It is known[10] that the solution to this problem exists and is unique up to the addition of a constant to p. With this choice of p, define $\mathbf{u} = \mathbf{w} - \operatorname{grad} p$.

[10] See R. Courant and D. Hilbert [1953], *Methods of Mathematical Physics,* Wiley. The equation $\Delta p = f$, $\partial p/\partial n = g$ has a solution unique up to a constant if and only if $\int_D f dV = \int_{\partial D} g\, dA$. The divergence theorem ensures that this condition is satisfied in our case.

Then, clearly $\mathbf{u}$ has the desired properties $\operatorname{div}\mathbf{u} = 0$, and also $\mathbf{u}\cdot\mathbf{n} = 0$ by construction of p. ■

The situation is shown schematically in Figure 1.3.2.

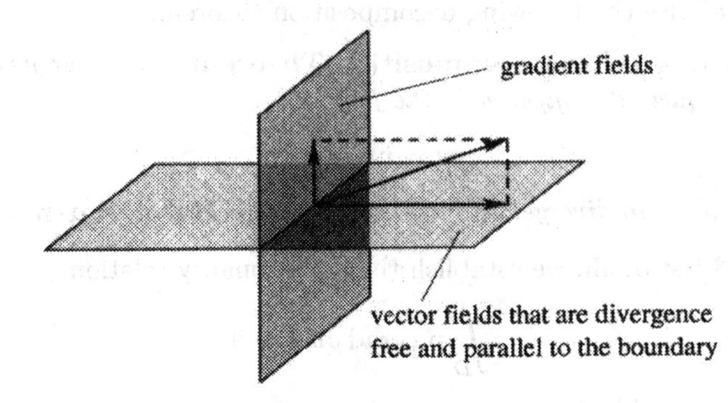

FIGURE 1.3.2. Decomposing a vector field into a divergence-free and gradient part.

It is natural to introduce the operator $\mathbb{P}$, an orthogonal projection operator, which maps $\mathbf{w}$ onto its divergence-free part $\mathbf{u}$. By the preceding theorem, $\mathbb{P}$ is well defined. Notice that by construction $\mathbb{P}$ is a linear operator and that

$$\mathbf{w} = \mathbb{P}\mathbf{w} + \operatorname{grad} p. \tag{1.3.7}$$

Also notice that

$$\mathbb{P}\mathbf{u} = \mathbf{u}$$

provided $\operatorname{div}\mathbf{u} = 0$ and $\mathbf{u}\cdot\mathbf{n} = 0$, and that

$$\mathbb{P}(\operatorname{grad} p) = 0.$$

Now we apply these ideas to the incompressible Navier–Stokes equations (1.3.5). If we apply the operator $\mathbb{P}$ to both sides, we obtain

$$\mathbb{P}(\partial_t\mathbf{u} + \operatorname{grad} p) = \mathbb{P}\left(-(\mathbf{u}\cdot\nabla)\mathbf{u} + \frac{1}{R}\Delta\mathbf{u}\right).$$

Because $\mathbf{u}$ is divergence-free and vanishes on the boundary, the same is true of $\partial_t\mathbf{u}$ (if $\mathbf{u}$ is smooth enough). Thus, by (1.3.7), $\mathbb{P}\partial_t\mathbf{u} = \partial_t\mathbf{u}$. Because $\mathbb{P}(\operatorname{grad} p) = 0$, we get

$$\partial_t\mathbf{u} = \mathbb{P}\left(-\mathbf{u}\cdot\nabla\mathbf{u} + \frac{1}{R}\Delta\mathbf{u}\right). \tag{1.3.8}$$

Although $\Delta\mathbf{u}$ is divergence free, it need not be parallel to the boundary and so we *cannot* simply write $\mathbb{P}\Delta\mathbf{u} = \mathbf{0}$. This form (1.3.8) of the Navier–Stokes equations eliminates the pressure and expresses $\partial_t\mathbf{u}$ in terms of $\mathbf{u}$ alone. The pressure can then be recovered as the gradient part of

$$-\mathbf{u}\cdot\nabla\mathbf{u} + \frac{1}{R}\Delta\mathbf{u}.$$

This form (1.3.8) of the equations is not only of theoretical interest, shedding light on the role of the pressure, but is of practical interest for numerical algorithms.[11]

The pressure in compressible flows is conceptually different than in incompressible flows just as it was in ideal flow. If we think of viscous flow as ideal flow with viscous effects added on, it is not unreasonable to assume that p is still a function of ρ.

A note of caution is appropriate here. The expressions for $p(\rho)$ used in practical situations are often borrowed from the science of equilibrium thermodynamics. It is not obvious that p as defined here (through equation (1.3.1)) is identical to p as defined in that other science. Not all quantities called p are equal. The use of expressions from equilibrium thermodynamics requires an additional physical justification, which is indeed often available, but which should not be forgotten.

According to the analysis given earlier, the pressure p in incompressible flow is determined by the equation of continuity $\operatorname{div}\mathbf{u} = 0$. To see why this is physically reasonable, consider a compressible flow with $p = p(\rho)$, where $p'(\rho) > 0$. If fluid flows into a given fixed volume V, the density in V will increase, and if $p'(\rho) > 0$, then p in V will also increase. If either the change in ρ is large enough or $p'(\rho)$ is large enough, $-\operatorname{grad} p$ at the boundary of V will begin to point away from V, and through the term $-\operatorname{grad} p$ in the equation for $\partial_t\mathbf{u}$, this will cause the fluid to flow away from V. Thus, the pressure controls and moderates the variations in density. If the density is to remain invariant, this must be accomplished by an appropriate p, that is, $\operatorname{div}\mathbf{u} = 0$ determines p.

In the Navier–Stokes equations for a viscous incompressible fluid, namely,

$$\partial_t\mathbf{u} + (\mathbf{u}\cdot\nabla)\mathbf{u} = -\nabla p + \frac{1}{R}\Delta\mathbf{u},$$

we call

$$\frac{1}{R}\Delta\mathbf{u}, \quad \text{the } \textit{\textbf{diffusion}} \text{ or } \textit{\textbf{dissipation}} \text{ term,}$$

[11] See, for instance, A. J. Chorin, *Math. Comp.* **23** [1969], 341-353 for algorithms, and D. Ebin and J. E. Marsden, *Ann. of Math.* **92** [1970], 102–163 for a theoretical investigation of the projection operator and the use of material coordinates.

and

$$(\mathbf{u} \cdot \nabla)\mathbf{u}, \quad \text{the } \textbf{\textit{inertia}} \text{ or } \textbf{\textit{convective}} \text{ term.}$$

The equations say that $\mathbf{u}$ is convected subject to pressure forces and, at the same time, is dissipated. Suppose R is very small. If we write the equations in the form $\partial_t \mathbf{u} = \mathbb{P}(-\mathbf{u} \cdot \nabla \mathbf{u} + \frac{1}{R}\Delta \mathbf{u})$, we see that they are approximated by

$$\partial_t \mathbf{u} = \mathbb{P}\left(\frac{1}{R}\Delta \mathbf{u}\right),$$

that is,

$$\partial_t \mathbf{u} = -\operatorname{grad} p + \frac{1}{R}\Delta \mathbf{u} \qquad \text{and} \qquad \operatorname{div} \mathbf{u} = 0,$$

which are the ***Stokes' equations*** for incompressible flow. These are linear equations of "parabolic" type. For small R (i.e., slow velocity, large viscosity, or small bodies), the solution of the Stokes equation provides a good approximation to the solution of the Navier–Stokes equations. Later, we shall mostly be interested in flows with the large R; for these the inertial term is important and in some sense is dominant. We hesitate and say "in some sense" because no matter how small $(1/R)\Delta \mathbf{u}$ may be, it still produces a large effect, namely, the change in boundary conditions from $\mathbf{u} \cdot \mathbf{n} = 0$ when $(1/R)\Delta \mathbf{u}$ is absent to $\mathbf{u} = \mathbf{0}$ when it is present.

There is a major difference between the ideal and viscous flow with regard to the energy of the fluid. The viscous terms provide a mechanism by which macroscopic energy can be converted into internal energy. General principles of thermodynamics state that this energy transfer is one-way. In particular, for incompressible flow, we should have

$$\frac{d}{dt} E_{\text{kinetic}} \leq 0. \tag{1.3.9}$$

We calculate $(d/dt)E_{\text{kinetic}}$ for incompressible viscous flow using the transport theorem, as we did in §**1.1**. We get

$$\begin{aligned} \frac{d}{dt} E_{\text{kinetic}} &= \frac{d}{dt} \frac{1}{2} \int_D \rho \|\mathbf{u}\|^2 dV = \int_D \rho \mathbf{u} \cdot \frac{D\mathbf{u}}{Dt} \, dV \\ &= \int_D \left(-\mathbf{u} \cdot \nabla p + \frac{1}{R} \mathbf{u} \cdot \Delta \mathbf{u} \right) dV, \end{aligned}$$

by (1.3.3) and $\operatorname{div} \mathbf{u} = 0$. Because $\mathbf{u}$ is orthogonal to $\operatorname{grad} p$, we get

$$\frac{d}{dt} E_{\text{kinetic}} = \frac{1}{R} \int_D \mathbf{u} \cdot \Delta \mathbf{u} \, dV.$$

The vector identity $\operatorname{div}(f\mathbf{V}) = f \operatorname{div} \mathbf{V} + \mathbf{V} \cdot \nabla f$ gives

$$\nabla \cdot (u\nabla u + v\nabla v + w\nabla w) \\ = \nabla u \cdot \nabla u + \nabla v \cdot \nabla v + \nabla w \cdot \nabla w + u\Delta u + v\Delta v + w\Delta w.$$

This equation, the divergence theorem, and the boundary condition $\mathbf{u} = \mathbf{0}$ on ∂D enable us to simplify the expression for $(d/dt)E_{\text{kinetic}}$ to

$$\frac{d}{dt}E_{\text{kinetic}} = -\mu \int_D \|\nabla \mathbf{u}\|^2 \, dV, \tag{1.3.10}$$

where $\|\nabla \mathbf{u}\|^2 = \nabla \mathbf{u} \cdot \nabla \mathbf{u} = \|\nabla u\|^2 + \|\nabla v\|^2 + \|\nabla w\|^2$. Notice that (1.3.9) and (1.3.10) are compatible exactly when $\mu \geq 0$ (or, equivalently, $\nu \geq 0$ or $0 < R \leq \infty$). In other words, there is no such thing as "negative viscosity."

A similar analysis for compressible flow and making use of (1.3.2)' leads to the inequalities

$$\mu \geq 0 \quad \text{and} \quad \lambda + \tfrac{2}{3}\mu \geq 0$$

and with $\boldsymbol{\sigma}$ given by (1.3.2).[12]

At the end of §**1.1** we noted that ideal flow in a channel leads to unreasonable results. We now reconsider this example with viscous effects.

Example Consider stationary viscous incompressible flow between two stationary plates located at $y = 0$ and $y = 1$, as shown in Figure 1.3.3. We seek a solution for which $\mathbf{u}(x, y) = (u(x, y), 0)$ and p is only a function of x, with $p_1 = p(0), p_2 = p(L)$, and $p_1 > p_2$, so the fluid is "pushed" in the positive x direction. The incompressible Navier–Stokes equations are

$$\partial_x u = 0 \qquad \text{(incompressibility)}$$

and

$$0 = -u\,\partial_x u - \partial_x p + \frac{1}{R}\left[\partial_x^2 u + \partial_y^2 u\right]$$

with boundary conditions $u(x, 0) = u(x, 1) = 0$. Because $\partial_x u = 0, u$ is only a function of y and thus, writing $u(x, y) = u(y)$, we obtain

$$p' = \frac{1}{R}u''.$$

Because each side depends on different variables,

$$p' = \text{constant}, \quad \frac{1}{R}u'' = \text{constant}.$$

[12] See, for example, S. Chapman and T. G. Cowling, *The Mathematical Theory of Non-uniform Gases*, Cambridge University Press, 1958.

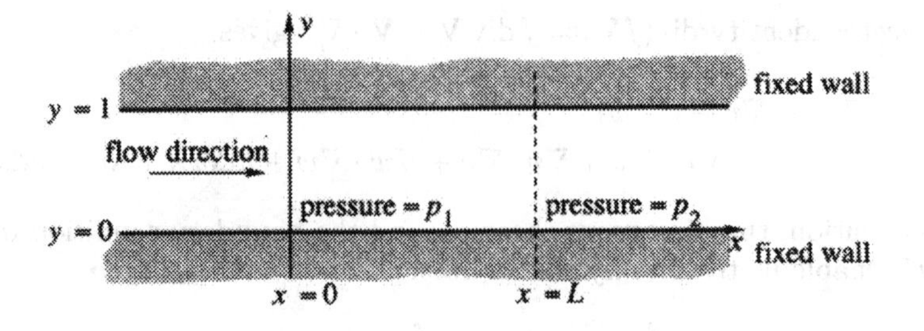

FIGURE 1.3.3. Flow between two parallel plates; the fluid is pushed from left to right and correspondingly, $p_1 > p_2$.

Integration gives

$$p(x) = p_1 - \frac{\Delta p}{L} x, \qquad \Delta p = p_1 - p_2,$$

and

$$u(y) = y(1-y) R \frac{\Delta p}{2L}.$$

Notice that the velocity profile is a parabola (Figure 1.3.4).

The presence of viscosity allows the pressure forces to be balanced by the term $\frac{1}{R}u''(y)$ and allows the fluid to achieve a stationary state. We saw that this was not possible for ideal flow. ♦

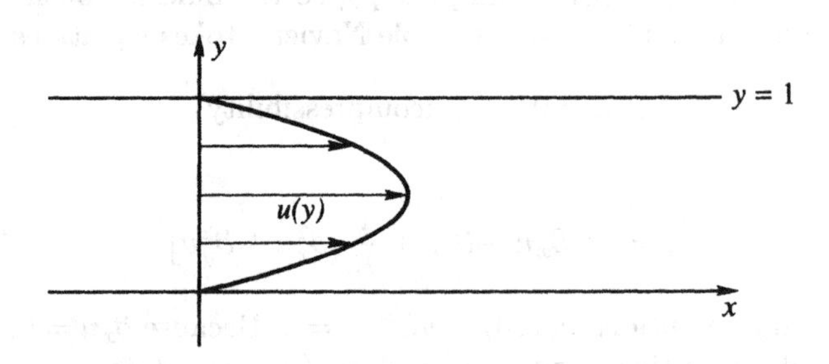

FIGURE 1.3.4. Viscous flow between two plates.

Next we consider the vorticity equation for (homogeneous) viscous incompressible flow. In the two-dimensional case we proved in §**1.2** (see equation (1.2.12)) that for isentropic ideal plane flow, $D\xi/Dt = 0$. The derivation is readily modified to cover viscous incompressible flow; the result is

$$\frac{D\xi}{Dt} = \frac{1}{R}\Delta\xi. \tag{1.3.11}$$

This shows that the vorticity is diffused by viscosity as well as being tranported by the flow. Introduce the stream function $\psi(x,y,t)$ by means of $(1.2.15)_2$ and $(1.2.15)_3$ as before. We saw that we could impose the boundary condition $\psi = 0$ on ∂D. Now, however, the no-slip condition $\mathbf{u} = \mathbf{0}$ on ∂D implies that

$$\partial_x \psi = 0 = \partial_y \psi \quad \text{on } \partial D$$

by $(1.2.15)_3$. Because $\psi = 0$ on ∂D implies that the tangential derivative of ψ on ∂D vanishes, we get the extra boundary condition

$$\frac{\partial \psi}{\partial \mathbf{n}} = 0 \quad \text{on } \partial D$$

This extra condition should be somewhat mystifying; certainly we cannot impose it when we solve $\Delta \psi = -\xi, \psi = 0$ on ∂D, because this problem already has a solution. Thus, it is not clear how to get the system

$$\left.\begin{aligned} \frac{D\xi}{Dt} &= \frac{1}{R}\Delta\xi, \\ \Delta\psi = -\xi, &\qquad \psi = 0 \quad \text{on } \partial D, \\ u = \partial_y \psi, &\qquad v = -\partial_x \psi \end{aligned}\right\} \tag{1.3.12}$$

to work. We shall study this problem in §**2.2**.

For three-dimensional viscous incompressible flow, the vorticity equation is

$$\frac{D\boldsymbol{\xi}}{Dt} - (\boldsymbol{\xi}\cdot\nabla)\mathbf{u} = \frac{1}{R}\Delta\boldsymbol{\xi}. \tag{1.3.13}$$

Thus, vorticity is convected, stretched, and diffused. (The left-hand side of (1.3.13) is called the ***Lie derivative***. It is this *combination*, rather than each term *separately*, that makes coordinate independent sense.) Here the problems with getting a system like (1.3.12) are even worse; even in the isentropic case we had trouble with (1.2.16) because of boundary conditions.

For viscous flow, circulation is no longer a constant of the motion. One might suspect from (1.3.13) that if $\boldsymbol{\xi} = \mathbf{0}$ at $t = 0$, then $\boldsymbol{\xi} = \mathbf{0}$ for all time. However, this is not true: viscous flow allows for the generation of vorticity. This is possible because of the difference in boundary conditions between ideal and viscous flows. The mechanism of vorticity generation is related to the difficulties with the boundary conditions in equations (1.3.12) and will be discussed in §**2.2**.

For many of our discussions we have made the assumption of incompressibility. We now give a heuristic analysis of when such an assumption will be reasonable and when, instead, the compressible equations should be used. We shall do this in the context of isentropic stationary flows for simplicity. Assume that we have an ***equation of state***

$$p = p(\rho), \qquad p'(\rho) > 0.$$

Define

$$c = \sqrt{p'(\rho)}.$$

For reasons that will become clear later, c is called the ***sound speed*** of the fluid. Thus, we have

$$c^2 d\rho = dp. \tag{1.3.14}$$

Let $u = \|\mathbf{u}\|$ be the flow speed. One calls $M = u/c$ the (local) ***Mach number*** of the flow; it is a function of position in the flow. From Bernoulli's theorem proved in §**1.1**,

$$\frac{u^2}{2} + \int \frac{dp}{\rho(p)} = \text{constant on streamlines.} \tag{1.3.15}$$

Also, differentiating the continuity equation in the form (1.2.10) along streamlines gives

$$0 = J d\rho + \rho\, dJ, \tag{1.3.16}$$

where J is the Jacobian of the flow map. Combining (1.3.14), (1.3.15), and (1.3.16) we get

$$\frac{dJ}{J} = -M\frac{du}{c}.$$

The flow will be approximately incompressible if J changes only by a small amount along streamlines. Thus, a steady flow can be viewed as incompressible if the flow speed is much less than the sound speed,

$$u \ll c, \quad \text{i.e.,} \quad M \ll 1,$$

or if changes in the speed along streamlines are very small compared to the sound speed.

For example, for equations of state of the kind associated with ideal gases,

$$p = A\rho^\gamma, \qquad \gamma > 1,$$

we have

$$c = \sqrt{\frac{\gamma p}{\rho}}$$

so the flow will be approximately incompressible if γ is very large.

For nonsteady flow one also needs to know that

$$\frac{l}{\tau} \ll c,$$

where l is a characteristic length and τ is a characteristic time over which the flow picture changes appreciably.[13] The presence of viscosity does not alter these conclusions significantly.

Exercises

◇ **Exercise 1.3-1** Find a stationary viscous incompressible flow in a circular pipe with radius $a > 0$ and with pressure gradient ∇p.

◇ **Exercise 1.3-2** Show that the incompressible Navier–Stokes equations in cylindrical coordinates are

(i) $\rho\left(\frac{Dv_r}{Dt} - \frac{v_\theta^2}{r}\right) = \rho f_r - \frac{\partial p}{\partial r} + \mu\left(\Delta v_r - \frac{v_r}{r^2} - \frac{2}{r^2}\frac{\partial v_\theta}{\partial_\theta}\right).$

(ii) $\rho\left(\frac{Dv_\theta}{Dt} + \frac{v_r v_\theta}{r}\right) = \rho f_\theta - \frac{1\partial p}{r\partial\theta} + \mu\left(\Delta v_\theta + \frac{2\partial v_2}{r^2\partial\theta} - \frac{v_\theta}{r^2}\right).$

(iii) $\rho\frac{Dv_z}{Dt} = \rho f_z - \frac{\partial p}{\partial z} + \mu\Delta v_z,$

where
$$\Delta = \frac{1}{r}\frac{\partial}{\partial r}\left(r\frac{\partial}{\partial r}\right) + \frac{1}{r^2}\frac{\partial^2}{\partial\theta^2} + \frac{\partial^2}{\partial z^2}$$

and
$$\frac{D}{Dt} = \frac{\partial}{\partial t} + v_r\frac{\partial}{\partial r} + \frac{v_\theta}{r}\frac{\partial}{\partial\theta} + v_z\frac{\partial}{\partial z}.$$

◇ **Exercise 1.3-3 Flow in an infinite pipe.**

(i) ***Poiseuille flow***. Work in cylindrical coordinates with a pipe of radius a aligned along the z-axis. The no-slip boundary condition is $\mathbf{v} = \mathbf{0}$ when $r = a$. Assume the solution takes the form $p = Cz$, C constant, $v_z = v_z(r)$, and $v_r = v_\theta = 0$. Using Exercise 1.3-2, obtain

$$C = \mu\Delta v_z = \mu\left(\frac{1}{r}\frac{\partial}{\partial r}\left(r\frac{\partial v_z}{\partial r}\right)\right).$$

[13]Theoretical work on the limit $c \to \infty$ is given by D. Ebin, *Ann. Math.* **141** [1977], 105, and S. Klainerman and A. Majda, *Comm. Pure Appl. Math.*, **35** [1982], 629. Algorithms for solving the equations for incompressible flow by exploiting the regularity of the limit $c \to \infty$ can be found in A. J. Chorin, *J. Comp. Phys.* **12** [1967], 1.

Integration yields

$$v_z = -\frac{C}{4\mu} r^2 + A \log r + B,$$

where A, B are constants. Because we require that the solution be bounded, A must be 0, because $\log r \to -\infty$ as $r \to 0$. Use the no-slip condition to determine B and obtain

$$v_z = \frac{C}{4\mu}(a^2 - r^2).$$

(ii) Show that the ***mass flow rate*** $Q = \int_s \rho v_z \, dA$ through the pipe is $Q = \rho\pi C a^4 / 8\mu$. This is the so-called ***fourth-power law.***

(iii) Determine the pressure on the walls.

◇ **Exercise 1.3-4** Compute the solution to the problem of stationary viscous flow between two concentric cylinders and determine the pressure on the walls. (Hint: Proceed as above, but retain the log term.)

2 Potential Flow and Slightly Viscous Flow

The goal of this chapter is to present a deeper study of the relationship between viscous and nonviscous flows. We begin with a more detailed study of inviscid irrotational flows, that is, potential flows. Then we go on to study boundary layers, where the main difference between slightly viscous and inviscid flows originates.

This is further developed in the third section using probabilistic methods. For most of this chapter we will study incompressible flows. A detailed study of some special compressible flows is the subject of Chapter **3**.

2.1 Potential Flow

Throughout this section, all flows are understood to be ideal (*i.e.*, inviscid); in other words, either incompressible and nonviscous or isentropic and nonviscous. Although we allow both, our main emphasis will be on the incompressible case.

A flow with zero vorticity is called ***irrotational***. For ideal flow, this holds for all time if it holds at one time by the results of §**1.2**. An inviscid, irrotational flow is called a ***potential flow***. A domain D is called ***simply connected*** if any continuous closed curve in D can be continuously shrunk to a point without leaving D. For example, in space, the exterior of a solid sphere is simply connected, whereas in the plane the exterior of a solid disc is not simply connected.

For irrotational flow in a simply connected region D, there is a scalar function $\varphi(x,t)$ on D called the ***velocity potential*** such that for each t, $\mathbf{u} = \operatorname{grad}\varphi$. In particular, it follows that the circulation around any closed curve C in D is zero. Using the identity

$$(\mathbf{u}\cdot\nabla)\mathbf{u} = \tfrac{1}{2}\nabla\left(\|\mathbf{u}\|^2\right) - \mathbf{u}\times(\nabla\times\mathbf{u}), \tag{2.1.1}$$

we can write the equations for isentropic potential flow in the form

$$\partial_t\mathbf{u} + \tfrac{1}{2}\nabla(\|\mathbf{u}\|^2) = -\operatorname{grad} w,$$

where w is the enthalpy, as in §**1.1**. Substituting $\mathbf{u} = \operatorname{grad}\varphi$, we obtain

$$\operatorname{grad}\left(\partial_t\varphi + \tfrac{1}{2}\|\mathbf{u}\|^2 + w\right) = 0,$$

and thus

$$\partial_t\varphi + \tfrac{1}{2}\|\mathbf{u}\|^2 + w = \text{ constant in space.} \tag{2.1.2}$$

In particular, if the flow is stationary,

$$\tfrac{1}{2}\|\mathbf{u}\|^2 + w = \text{ constant in space.}$$

For the last equation to hold, simple connectivity of D is unnecessary. The version of Bernoulli's theorem given in §**1.1** concluded that $\frac{1}{2}\|\mathbf{u}\|^2 + w$ was constant on streamlines. The stronger conclusion here results from the additional irrotational hypothesis, $\boldsymbol{\xi} = \mathbf{0}$. For homogeneous incompressible ideal flow, note that $w = p/\rho_0$ from the definition of w.

For potential flow in nonsimply connected domains, it can occur that the circulation Γ_C around a closed curve C is nonzero. For instance, consider

$$\mathbf{u} = \left(\frac{-y}{x^2+y^2}, \frac{x}{x^2+y^2}\right)$$

outside the origin. If the contour C can be deformed within D to another contour C', then $\Gamma_C = \Gamma_{C'}$; see Figure 2.1.1.

The reason is that basically $C \cup C'$ forms the boundary of a surface Σ in D. Stokes' theorem then gives

$$\int_\Sigma \boldsymbol{\xi}\cdot d\mathbf{A} = \int_C \mathbf{u}\cdot ds - \int_{C'} \mathbf{u}\cdot ds = \Gamma_C - \Gamma_{C'}$$

and because $\boldsymbol{\xi} = \mathbf{0}$ in D, we get $\Gamma_C = \Gamma_{C'}$. (A more careful argument proving the invariance of Γ_C under deformation is given in books on complex variables.) Notice that from §**1.2**, the circulation around a curve is constant in time. Thus, the circulation around an obstacle in the plane is well-defined and is constant in time.

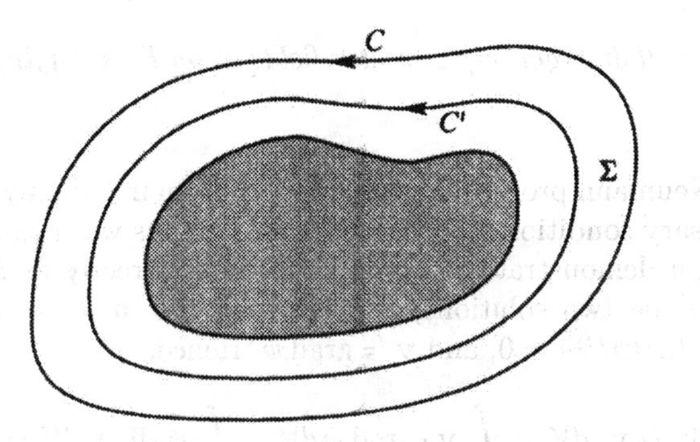

FIGURE 2.1.1. The circulations about C and C' are equal if the flow is potential in Σ.

Next, consider incompressible potential flow in a simply connected domain D. From $\mathbf{u} = \operatorname{grad} \varphi$ and $\operatorname{div} \mathbf{u} = 0$, we have

$$\Delta\varphi = 0.$$

Let the velocity of ∂D be specified as $\mathbf{V}$, so

$$\mathbf{u} \cdot \mathbf{n} = \mathbf{V} \cdot \mathbf{n}.$$

Thus, φ solves the Neumann problem:

$$\Delta\varphi = 0, \qquad \frac{\partial\varphi}{\partial n} = \mathbf{V} \cdot \mathbf{n}. \tag{2.1.3}$$

If φ is a solution, then $\mathbf{u} = \operatorname{grad} \varphi$ is a solution of the stationary homogeneous Euler equations, *i.e.*,

$$\begin{aligned} \rho(\mathbf{u} \cdot \nabla)\mathbf{u} &= -\operatorname{grad} p, \\ \operatorname{div} \mathbf{u} &= 0, \\ \mathbf{u} \cdot \mathbf{n} &= \mathbf{V} \cdot \mathbf{n} \quad \text{on } \partial D, \end{aligned} \tag{2.1.4}$$

where $p = -\rho\|\mathbf{u}\|^2/2$. This follows from the identity (2.1.1). Therefore, solutions of (2.1.3) are in one-to-one correspondence with irrotational solutions of (2.1.4) (with φ determined only up to an additive constant) on simply connected regions. This observation leads to the following.

Theorem *Let D be a simply connected, bounded region with prescribed velocity $\mathbf{V}$ on ∂D. Then*

i *there is exactly one potential homogeneous incompressible flow (satisfying (2.1.4)) in D if and only if $\int_{\partial D} \mathbf{V} \cdot \mathbf{n}\, dA = 0$;*

ii *this flow is the minimizer of the kinetic energy function*

$$E_{\text{kinetic}} = \frac{1}{2}\int_D \rho\,\|\mathbf{u}\|^2\, dV,$$

among all divergence-free vector fields $\mathbf{u}'$ *on* D *satisfying* $\mathbf{u}' \cdot \mathbf{n} = \mathbf{V} \cdot \mathbf{n}$.

Proof

i The Neumann problem (2.1.3) has a solution if and only if the obvious necessary condition $\int_{\partial D} \mathbf{V} \cdot \mathbf{n}\, dA = 0$ holds, as was mentioned earlier. We can demonstrate the uniqueness of $\mathbf{u}$ directly as follows: Let $\mathbf{u}$ and $\mathbf{u}'$ be two solutions, and let $\mathbf{v} = \mathbf{u} - \mathbf{u}', \psi = \varphi - \varphi'$. Then $\Delta\psi = 0, \partial\psi/\partial n = 0$, and $\mathbf{v} = \operatorname{grad}\psi$. Hence,

$$\int_D \operatorname{div}(\psi\mathbf{v})\, dV = \int_D \mathbf{v} \cdot \operatorname{grad}\psi\, dV + \int_D \psi \operatorname{div}\mathbf{v}\, dV = \int_D \mathbf{v} \cdot \mathbf{v}\, dV.$$

On the other hand,

$$\int_D \operatorname{div}(\psi\mathbf{v})\, dV = \int_{\partial D} \psi\mathbf{v} \cdot \mathbf{n}\, dA = 0.$$

Thus, $\int_D \|\mathbf{v}\|^2 dV = 0$ and $\mathbf{v} = 0$, that is, $\mathbf{u} = \mathbf{u}'$.

ii Let $\mathbf{u}$ solve (2.1.4) and let $\mathbf{u}'$ be divergence free and $\mathbf{u}' = \mathbf{n} = \mathbf{V} \cdot \mathbf{n}$. Let $\mathbf{v} = \mathbf{u} - \mathbf{u}'$; then $\operatorname{div}\mathbf{v} = 0$ and $\mathbf{v} \cdot \mathbf{n} = 0$ on ∂D. Therefore,

$$\begin{aligned} E_{\text{kinetic}} - E'_{\text{kinetic}} &= \tfrac{1}{2} \int_D \rho(\|\mathbf{u}\|^2 - \|\mathbf{u}'\|^2)\, dV \\ &= -\tfrac{1}{2} \int_D \rho\|\mathbf{u} - u'\|^2 dV + \int_D \rho(\mathbf{u} - u') \cdot \mathbf{u}\, dV \\ &\leq \int_D \rho\mathbf{v} \cdot \operatorname{grad}\varphi\, dV = 0. \end{aligned}$$

The last equality follows by the orthogonality relation proved in §**1.3**. Thus,

$$E_{\text{kinetic}} \leq E'_{\text{kinetic}}$$

as claimed. ■

Notice, in particular, that the only incompressible potential flow in a bounded region with fixed boundary is the trivial flow $\mathbf{u} = \mathbf{0}$. For unbounded regions this is not true without a careful specification of what can happen at infinity; the above uniqueness proof is valid only if the integration by parts (i.e., use of the divergence theorem) can be justified. For example, in polar coordinates in the plane,

$$\varphi(r, \theta) = \left(r + \frac{1}{r}\right) \cos\theta$$

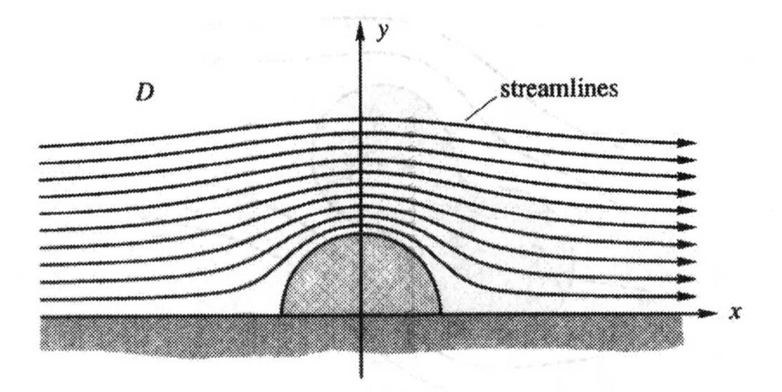

FIGURE 2.1.2. Potential flow in the upper half-plane outside the unit circle.

solves (2.1.3) with $\partial\varphi/\partial n = 0$ on the unit circle and on the x-axis. It represents a nontrivial irrotational potential flow on the simply connected region D shown in Figure 2.1.2. This flow may be arrived at by the methods of complex variables to which we will now turn.

Incompressible potential flow is very special, but is a key building block for understanding complicated flows. For plane flows the methods of complex variables are useful tools.

Let D be a region in the plane and suppose $\mathbf{u} = (u, v)$ is incompressible, that is,

$$\frac{\partial u}{\partial x} + \frac{\partial v}{\partial y} = 0 \tag{2.1.5}$$

and is irrotational, that is,

$$\frac{\partial u}{\partial y} - \frac{\partial v}{\partial x} = 0. \tag{2.1.6}$$

Let

$$F = u - iv, \tag{2.1.7}$$

which is called the ***complex velocity***. Equations (2.1.5) and (2.1.6) are exactly the Cauchy-Riemann equations for F, and so F is an analytic function on D. Conversely, given any analytic function F, $u = \operatorname{Re} F$ and $v = -\operatorname{Im} F$ define an incompressible (stationary) potential flow.

If F has a primitive, $F = dW/dz$, then we call W the ***complex potential***. (If one allows multivalued functions, W will always exist, but such a convention could cause confusion.) Write $W = \varphi + i\psi$. Then (2.1.7) is equivalent to

$$u = \partial_x\varphi = \partial_y\psi \quad \text{and} \quad v = \partial_y\varphi = -\partial_x\psi,$$

that is, $\mathbf{u} = \operatorname{grad}\varphi$ and ψ is the stream function. In what follows, however, we do not and must not assume a (single-valued) W exists.

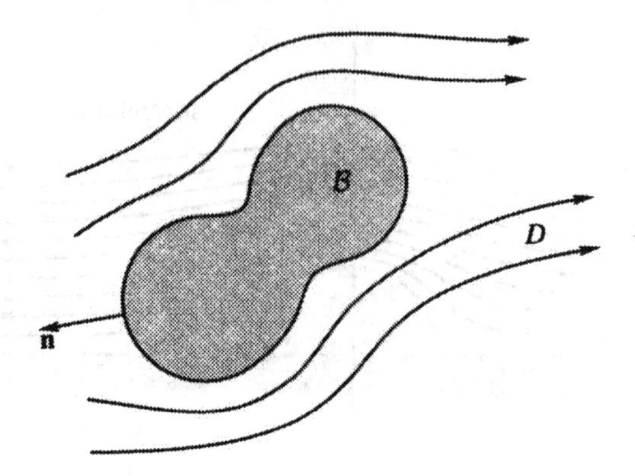

FIGURE 2.1.3. Flow around an obstacle.

Consider a flow in the exterior of an obstacle $\mathcal{B}$ (Figure 2.1.3).

The force on the body equals the force exerted on $\partial\mathcal{B}$ by the pressure, that is,

$$\mathcal{F} = -\int_{\partial\mathcal{B}} p\mathbf{n}\, ds, \tag{2.1.8}$$

which means that for any fixed vector $\mathbf{a}$,

$$\mathcal{F} \cdot \mathbf{a} = -\int_{\partial\mathcal{B}} p\mathbf{n} \cdot \mathbf{a}\, ds.$$

Formula (2.1.8) was already discussed at length in §**1.1**. We next prove a theorem that gives a convenient expression for $\mathcal{F}$.

Blasius' Theorem *For incompressible potential flow exterior to a body $\mathcal{B}$ (with rigid boundary) and complex velocity F, the force $\mathcal{F}$ on the body is given by*

$$\mathcal{F} = \frac{-i\rho}{2} \overline{\left[\int_{\partial\mathcal{B}} F^2\, dz\right]} \tag{2.1.9}$$

where the overbar denotes complex conjugation and where the vector $\mathcal{F}$ is identified with a complex number in the standard way; i.e., (x, y) is identified with $z = x + iy$.

Proof If $dz = dx + i\,dy$ represents an infinitesimal displacement along the boundary curve $C = \partial\mathcal{B}$, then $(1/i)dz = dy - i\,dx$ represents a normal displacement. Thus, by (2.1.8)

$$\mathcal{F} = -\int_C p\, dy + i\int_C p\, dx = i\int_C p(dx + i\, dy).$$

As we observed in (2.1.4),

$$p = \frac{-\rho(u^2+v^2)}{2}, \quad \text{and therefore} \quad \mathcal{F} = \frac{-i\rho}{2}\int_C (u^2+v^2)\,dz.$$

On the other hand, $F^2 = (u - iv)^2 = u^2 - v^2 - 2iuv$, and because $\mathbf{u}$ is parallel to the boundary, we get $u\,dy = v\,dx$. Thus,

$$F^2 dz = (u^2 - v^2 - 2iuv)(dx + i\,dy) = (u^2+v^2)(dx - i\,dy),$$

and because $u^2 + v^2$ is real, $\overline{F^2\,dz} = (u^2+v^2)\,dz$. ■

This formula will be used to prove the following (Figure 2.1.4):

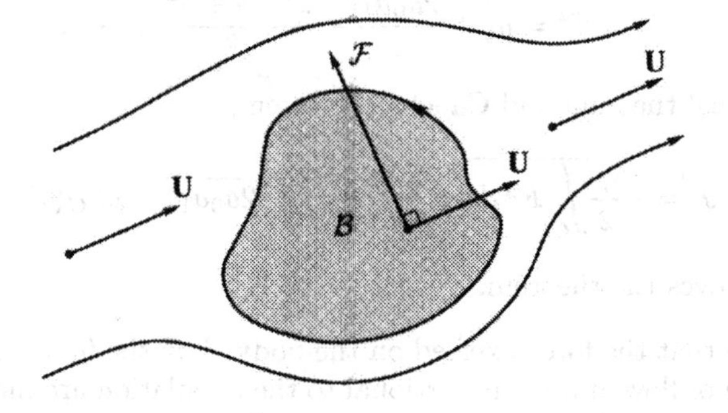

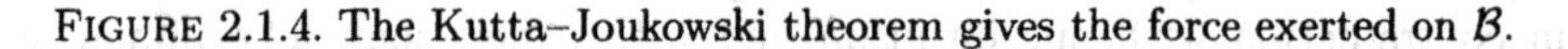

FIGURE 2.1.4. The Kutta–Joukowski theorem gives the force exerted on $\mathcal{B}$.

Kutta–Joukowski Theorem *Consider incompressible potential flow exterior to a region $\mathcal{B}$. Let the velocity field approach the constant value $(U, V) = \mathbf{U}$ at infinity. Then the force exerted on $\mathcal{B}$ is given by*

$$\mathcal{F} = -\rho\Gamma_C\|\mathbf{U}\|\mathbf{n}, \tag{2.1.10}$$

where Γ_C is the circulation around $\mathcal{B}$ and $\mathbf{n}$ is a unit vector orthogonal to $\mathbf{U}$.

Proof By assumption, the complex velocity F is an analytic function outside the body $\mathcal{B}$. It may therefore, be expanded in a Laurent series. Because F tends to a constant $\mathbf{U}$ at infinity, no positive powers of z can occur in the expansion. In other words, F has the form

$$F = a_0 + \frac{a_1}{z} + \frac{a_2}{z^2} + \frac{a_3}{z^3} + \cdots$$

valid outside any disc centered at the origin and containing $\mathcal{B}$. Because $\mathbf{U}$ is the velocity at infinity, $a_0 = U - iV$. By Cauchy's theorem,

$$\int_C F\,dz = 2\pi a_1 i,$$

where $C = \partial\mathcal{B}$. (The integral is unchanged if we change C to a circle of large radius.) However,

$$\int_C F\,dz = \int_C (u - iv)(dx + i\,dy) = \int_C u\,dx + v\,dy = \int_C \mathbf{u}\cdot d\mathbf{s} = \Gamma_C$$

because $u\,dy = v\,dx$, *i.e.*, $\mathbf{u}$ is parallel to $\partial\mathcal{B}$. Therefore,

$$a_1 = \frac{\Gamma_C}{2\pi i}.$$

Squaring F gives the Laurent expansion

$$F^2 = a_0^2 + \frac{2a_0a_1}{z} + \frac{2a_0a_2 + a_1^2}{z^2} + \cdots .$$

By Blasius' theorem and Cauchy's theorem,

$$\mathcal{F} = -\frac{i\rho}{2}\overline{\int_C F^2\,dz} = -\frac{i\rho}{2}\cdot\overline{(2\pi i\cdot 2a_0a_1)} = \rho\Gamma_C(V - iU)$$

which proves the theorem. ■

Notice that the force exerted on the body $\mathcal{B}$ by the flow is normal to the direction of flow and is proportional to the circulation around the body. In any case, the body experiences no drag (i.e., no force opposing the flow) in contradiction with our intuition and with experiment. The difficulty, of course, stems from the fact that we have neglected viscosity. (We shall remedy this in the succeeding two sections.) Even worse, if $\Gamma_C = 0$, there is no net force on the body at all, a fact hard to reconcile with our intuition even for ideal flow. This result is called ***d'Alembert's paradox***.

Example 1 For a complex number $\alpha = U - iV$, let $W(z) = \alpha z$. Thus, $F(x) = \alpha$, so the velocity field is $\mathbf{u} = (U, V)$. This is two-dimensional flow moving with constant velocity in the direction (U, V). ♦

Example 2 Let $\mathcal{B}$ be the disc of radius $a > 0$ centered at the origin in the complex plane, and let

$$W(z) = U\left(z + \frac{a^2}{z}\right) \tag{2.1.11}$$

for a positive constant U. The complex velocity is

$$F(z) = W'(z) = U\left(1 - \frac{a^2}{z^2}\right), \tag{2.1.12}$$

which approaches U at ∞. The velocity potential φ and the stream function ψ are determined by $W = \varphi + i\psi$. To verify that the flow is tangent to the circle $|z| = a$, we need only to show that ψ = constant when $|z| = a$. In fact, for $|z|^2 = z\bar{z} = a^2$, we have from (2.1.1),

$$W(z) = U(z + \bar{z}),$$

so W is real on $|z| = a$, that is, $\psi = 0$ on $|z| = a$. The flow is shown in Figure 2.1.5.

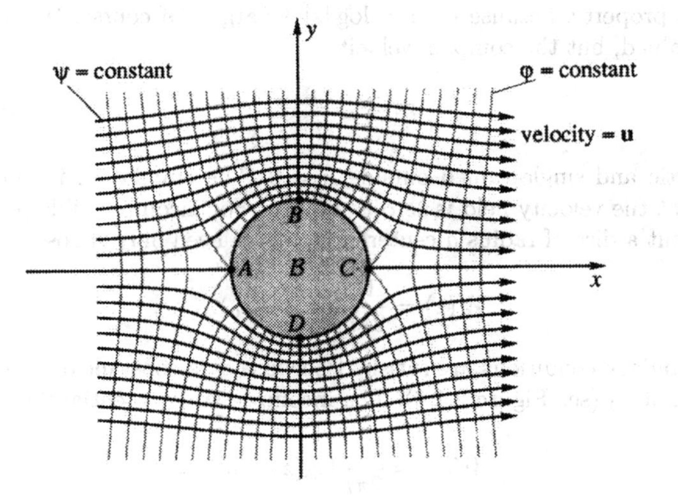

FIGURE 2.1.5. Potential flow around a disc.

From (2.1.12) with $z = ae^{i\theta}$, that is, z on $\partial\mathcal{B}$, we find

$$F(z) = U\left(1 - \frac{a^2}{a^2e^{2i\theta}}\right) = U(1 - \cos 2\theta + i \sin 2\theta).$$

Thus, the velocity is zero at A and C; that is, A and C are stagnation points. The velocity reaches a maximum at B and D. By Bernoulli's theorem, we can write

$$p = -\frac{\rho}{2}\|\mathbf{u}\|^2 + \text{constant};$$

thus, the pressure at A and C is maximum and is a minimum at B and D. The disc has zero circulation because $F = W'$ and W is single-valued.

If W is any analytic function defined in the whole plane, then

$$\tilde{W}(z) = W(z) + \overline{W\left(\frac{a^2}{\bar{z}}\right)}, \qquad |z| \geq a$$

is a potential describing a flow exterior to the disc of radius $a > 0$, but possibly with a more complicated behavior at infinity. This is proved along the same lines as in the argument just presented. ♦

Example 3 In §**1.2** we proved that choosing ψ to be an arbitrary increasing function of r alone yields a flow that is incompressible and has vorticity $\xi = -\Delta\psi$. If we can arrange for ψ to be the imaginary part of an analytic function, then the flow will be irrotational as well. The function

$$W(z) = \frac{\Gamma}{2\pi i}\log z \tag{2.1.13}$$

has this property, because $\log z = \log|z| + i\arg z$. Of course, $W(z)$ is not single-valued, but the complex velocity

$$F(z) = \frac{\Gamma}{2\pi i z} \tag{2.1.14}$$

is analytic and single-valued outside $z = 0$. The circulation is indeed Γ. Note that the velocity field is zero at infinity. For incompressible potential flow about a disc of radius a centered at z_0, we need only choose

$$W(z) = \frac{\Gamma}{2\pi i}\log(z - z_0).$$

The boundary conditions are satisfied because ψ is constant on any circle centered at z_0 (see Figure 2.1.6). The incompressible potential flow with

$$W(z) = \frac{\Gamma}{2\pi i}\log(z - z_0)$$

will be called a ***potential vortex*** at z_0. ♦

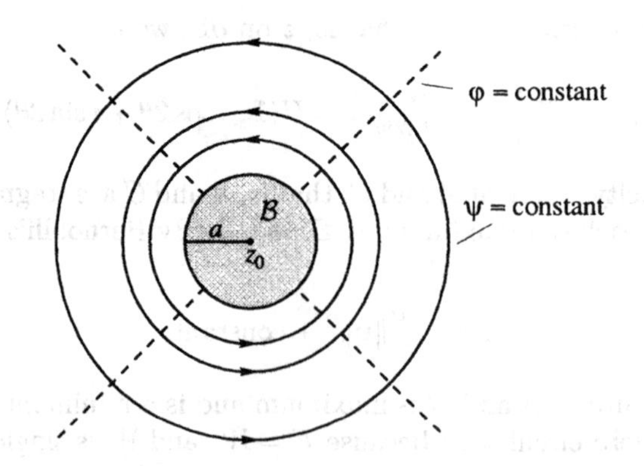

FIGURE 2.1.6. Potential vortex flow centered at z_0.

Example 4 We combine Examples 2 and 3 by forming

$$W(z) = U\left(z + \frac{a^2}{z}\right) + \frac{\Gamma}{2\pi i}\log z, \tag{2.1.15}$$

where $|z| \geq a$. Because ψ is constant on the boundary for each flow separately, it is also true for W given by (2.1.15). Thus, we get an incompressible potential flow on the exterior of the disc $|z| \leq a$ with circulation Γ around the disc. The velocity field is $(U, 0)$ at infinity (therefore, the Kutta–Joukowski theorem applies). On the surface of the disc the velocity $\mathbf{u} = \operatorname{grad} \varphi$ is tangent to the disc and is given in magnitude by

$$\text{velocity} = \frac{1}{r} \frac{\partial \varphi}{\partial \theta}\bigg|_{r=a}.$$

Here $\varphi = \operatorname{Re} W$, so that

$$\varphi(r, \theta) = U \cos\theta \left(r + \frac{a^2}{r} \right) + \frac{\Gamma\theta}{2\pi},$$

and thus

$$\text{velocity} = \frac{1}{a} \frac{\partial \varphi}{\partial \theta}\bigg|_{r=a} = -2U \sin\theta + \frac{\Gamma}{2\pi a}.$$

If $|\Gamma| < 4\pi a U$, there are two stagnation points A and C defined by

$$\sin\theta = \frac{\Gamma}{4\pi a U}$$

on the boundary, where the pressure is highest. See Figure 2.1.7.

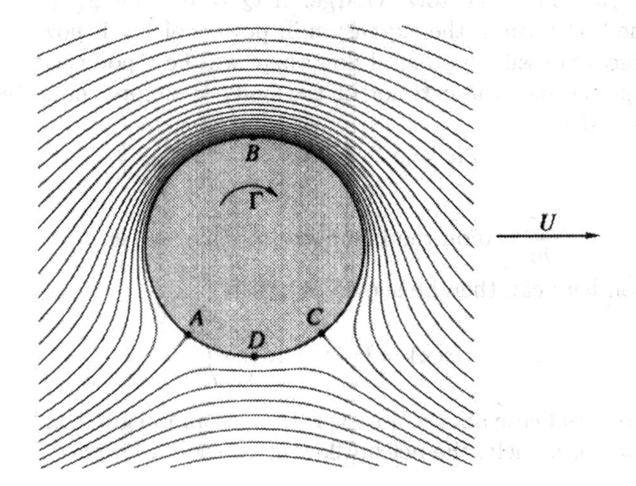

FIGURE 2.1.7. Flow around a disc with circulation.

This example helps to explain the Kutta–Joukowski theorem; note that the vertical lift may be attributed to the higher pressure at A and C. The symmetry in the y-axis means that there is no drag. ♦

D'Alembert's Paradox in Three Dimensions *In the case of steady incompressible potential flow around an obstacle in three dimensions with constant velocity U at infinity, not only can there be no drag, there can be no lift either.*

The difference with the two-dimensional case is explained by the fact that the exterior of a body in three-space is simply connected, whereas this is not true in two dimensions. We will not present the detailed proof of d'Alembert's paradox here, but we can give the idea.

Recall that the solution of $\Delta\varphi = -\rho$ in space is

$$\varphi(\mathbf{x}) = \frac{1}{4\pi}\int \frac{\rho(\mathbf{y})}{\|\mathbf{x}-\mathbf{y}\|}\, dV(\mathbf{y}),$$

that is, φ is the potential due to a charge distribution ρ. Notice that if ρ is concentrated in a finite region, then

$$\varphi(\mathbf{x}) = O\left(\frac{1}{r}\right),$$

where $r = \|\mathbf{x}\|$, that is,

$$|\varphi(\mathbf{x})| \leq \frac{\text{constant}}{r}$$

for large r. In fact, as we know physically, $\varphi(\mathbf{x}) \approx Q/4\pi r$ for r large, where $Q = \int \rho(\mathbf{y})\, dV(\mathbf{y})$ is the total charge. If $Q = 0$, then $\varphi(\mathbf{x}) = O(1/r^2)$ because the first term in the expansion in powers of $1/r$ is now missing.

For an incompressible potential flow there will be a potential φ, that is, $\mathbf{u} = \operatorname{grad}\varphi$ (because the exterior of the body is simply connected). The potential satisfies

$$\Delta\varphi = 0, \quad \nabla\varphi = \mathbf{U} \text{ at } \infty,$$

and

$$\frac{\partial\varphi}{\partial n} = 0 \text{ on the boundary of the obstacle.}$$

The solution here can then be shown to satisfy

$$\varphi(\mathbf{x}) = \mathbf{U}\cdot\mathbf{x} + O\left(\frac{1}{r}\right)$$

as in the potential case above. However, there is an integral condition analogous to $Q = 0$, namely, the net outflow at ∞ should be zero. This means

$$\varphi(\mathbf{x}) = \mathbf{U}\cdot\mathbf{x} + O\left(\frac{1}{r^2}\right).$$

Hence,

$$\mathbf{u} = \mathbf{U} + O(r^{-3}). \tag{2.1.16}$$

Because $p = -\rho v^2/2$, we also have $p = p_0 + O(r^{-3})$. (To see that this is true, write $\|\mathbf{u}\|^2 = U^2 + (\mathbf{u} - \mathbf{U}) \cdot (\mathbf{u} + \mathbf{U})$). The force on the body $\mathcal{B}$ is

$$\mathcal{F} = -\int_{\partial\mathcal{B}} p\mathbf{n}\, dA.$$

Let Σ be a surface containing $\mathcal{B}$. Because $\mathbf{u} \cdot \mathbf{n} = 0$ on $\partial\mathcal{B}$ and the flow is steady, equation (BM3) from **§1.1** applied to the region between $\mathcal{B}$ and Σ shows that

$$\mathcal{F} = -\int_{\Sigma} (\rho(\mathbf{u} \cdot \mathbf{n})\mathbf{u} + p\mathbf{n})\, dA.$$

We are free to choose Σ to be a sphere of large radius R enclosing the obstacle. Then

$$\begin{aligned} \mathcal{F} &= -\int_{\Sigma} (p_0\mathbf{n} + \rho(\mathbf{U} \cdot \mathbf{n})\mathbf{U})\, dA + (\text{area}\,\Sigma) \cdot O(R^{-3}) \\ &= \mathbf{0} + O(R^{-1}) \to 0 \text{ as } R \to \infty. \end{aligned}$$

Hence, $\mathcal{F} = \mathbf{0}$.

One may verify d'Alembert's paradox directly for flow past a sphere of radius $a > 0$. In this case

$$\varphi = -\frac{a^3}{2r^2}\mathbf{U} \cdot \mathbf{n} + \mathbf{x} \cdot \mathbf{U},$$

where $\mathbf{n} = \frac{\mathbf{x}}{\|\mathbf{x}\|}$, and

$$u = -\frac{a^3}{2r^2}\left[3\mathbf{n}(\mathbf{U} \cdot \mathbf{n}) - \mathbf{U}\right] + \mathbf{U},$$

where $\mathbf{U}$ is the velocity at infinity. We leave the detailed verification to the reader.[1] ♦

Next we will discuss a possible mechanism, ultimately to be justified by the presence of viscosity, by which one can avoid d'Alembert's paradox. An effort to resolve the paradox is of course prompted by the fact that real bodies in fluids do experience drag.

By an ***almost potential flow***, we mean a flow in which vorticity is concentrated in some thin layers of fluid; the flow is potential outside these thin layers, but there is a mechanism for producing vorticity near boundaries. For example, one can postulate that the flow past the obstacle shown in Figure 2.1.8 produces an almost potential flow with vorticity produced at the boundary and concentrated on two streamlines emanating from the body.

---

[1] See L. Landau and E. Lifschitz [1959] *Fluid Mechanics*, Pergamon, p. 34 for more information.

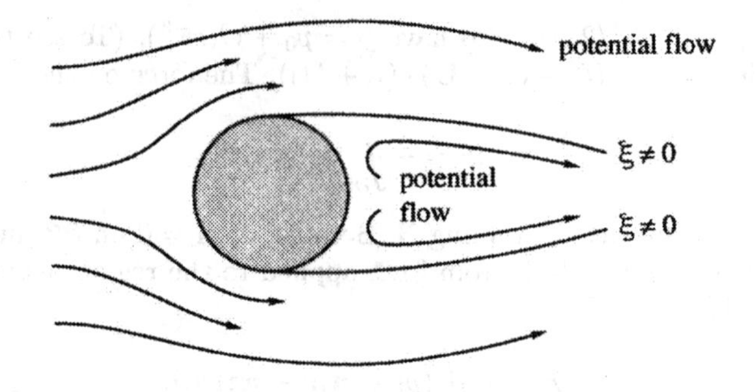

FIGURE 2.1.8. Almost potential flow has vorticity concentrated on two curves.

We image different potential flows in the two regions separated by these streamlines with the velocity field discontinuous across them. For such a model, the Kutta–Joukowski theorem does not apply and the drag may be different from zero. There are a number of situations in engineering where real flows can be usefully idealized as "nearly potential." These situations arise in particular when one considers "streamlined" bodies, that is, bodies so shaped as to reduce their drag. The discussion of such bodies, their design, and the validity of potential approximation to the flow around them are outside the scope of this book.

Next we shall examine a model of incompressible inviscid flow inspired by the idea of an almost potential flow and Example 3.

We imagine the vorticity in a fluid is concentrated in N ***vortices*** (i.e., points at which the vorticity field is singular), located at $\mathbf{x}_1, \mathbf{x}_2, \ldots, \mathbf{x}_N$ in the plane (Figure 2.1.9). The stream function of the jth vortex, ignoring the other vortices for a moment, is by Example 3,

$$\psi_j(\mathbf{x}) = -\frac{\Gamma_j}{2\pi} \log \|\mathbf{x} - \mathbf{x}_j\|. \tag{2.1.17}$$

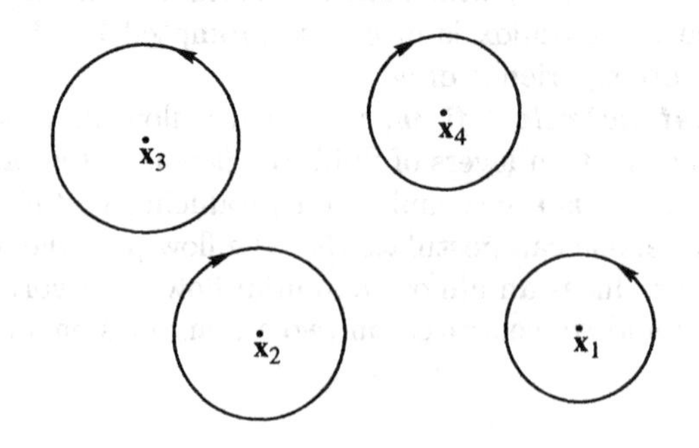

FIGURE 2.1.9. The flow generated by point vortices in the plane.

As the fluid moves according to Euler's equations, the circulations Γ_j associated with each vortex will remain constant. The vorticity field produced by the jth vortex can be written as

$$\xi_j = -\Delta\psi_j = \Gamma_j \delta(\mathbf{x} - \mathbf{x}_j),$$

where δ is the Dirac δ function. This equation arises from the fact, which we just accept, that the ***Green's function*** for the Laplacian in the plane is

$$G(\mathbf{x}, \mathbf{x}') = \frac{1}{2\pi} \log \|\mathbf{x} - \mathbf{x}'\|,$$

that is, G satisfies

$$\Delta_{\mathbf{x}} G(\mathbf{x}, \mathbf{x}') = \delta(\mathbf{x} - \mathbf{x}').$$

The solution of $\Delta\psi = -\xi$ is then given by the superposition

$$\psi(\mathbf{x}) = -\int \xi(\mathbf{x}') G(\mathbf{x} - \mathbf{x}')\, d\mathbf{x}'$$

which in our case reduces to $\psi(\mathbf{x}) = \sum_{j=1}^{N} \psi_j(\mathbf{x})$, where

$$\psi_j(\mathbf{x}) = -\frac{1}{2\pi} \Gamma_j \log \|\mathbf{x} - \mathbf{x}_j\|.$$

The velocity field induced by the jth vortex (again ignoring the other vortices) is given by

$$\mathbf{u}_j = (\partial_y \psi_j, -\partial_x \psi_j) = \left(-\frac{\Gamma_j}{2\pi} \left(\frac{y - y_j}{r^2} \right), \frac{\Gamma_j}{2\pi} \left(\frac{x - x_j}{r^2} \right) \right), \tag{2.1.18}$$

where $r = \|\mathbf{x} - \mathbf{x}_j\|$. Let the vortices all move according to the velocity field

$$\mathbf{u}(\mathbf{x}, t) = \sum_{j=1}^{N} \mathbf{u}_j(\mathbf{x}, t),$$

where $\mathbf{u}_j$ is given by (2.1.18) except we now allow, as we must, the centers of the vortices $\mathbf{x}_j, j = 1, \ldots, N$ to move. Each one ought to move as if convected by the net velocity field of the other vortices. Therefore, by (2.1.18), $\mathbf{x}_j$ moves according to the equations

$$\frac{dx_j}{dt} = -\frac{1}{2\pi} \sum_{i \neq j} \frac{\Gamma_i (y_j - y_i)}{r_{ij}^2} \quad \text{and} \quad \frac{dy_j}{dt} = \frac{1}{2\pi} \sum_{i \neq j} \frac{\Gamma_i (x_j - x_i)}{r_{ij}^2}, \tag{2.1.19}$$

where $r_{ij} = \|\mathbf{x}_i - \mathbf{x}_j\|$.

Let us summarize the construction of the flows we are considering: choose constants $\Gamma_1, \dots, \Gamma_N$ and initial points $\mathbf{x}_1 = (x_1, y_1), \dots, \mathbf{x}_N = (x_N, y_N)$ in the plane. Let these points move according to the equations (2.1.19). Define $\mathbf{u}_j$ by (2.1.18) and let

$$\mathbf{u}(\mathbf{x}, t) = -\sum_{j=1}^{N} \mathbf{u}_j(\mathbf{x}, t).$$

This construction produces formal solutions of Euler's equation ("formal" because the meaning of δ-function solutions of nonlinear equations is not obvious). These solutions have the property that the circulation theorem is satisfied. If C is a contour containing l vortices at $\mathbf{x}_1, \mathbf{x}_2, \dots, \mathbf{x}_l$, then $\Gamma_C = -\sum_{i=1}^{l} \Gamma_i$ and Γ_C is invariant under the flow. However, the relationship between these solutions and bona fide solutions of Euler's equations is not readily apparent. Such a relationship can, however, be established rigorously and such vortex systems do contain significant information about the solutions of Euler's equations under a wide variety of conditions.[2]

An important property of the equations is that they form a ***Hamiltonian system***. Define

$$H = -\frac{1}{4\pi} \sum_{i \neq j} \Gamma_i \Gamma_j \log \|\mathbf{x}_i - \mathbf{x}_j\|. \tag{2.1.20}$$

First of all, it is easy to check that (2.1.19) is equivalent to

$$\Gamma_j \frac{dx_j}{dt} = \frac{\partial H}{\partial y_j}, \quad \Gamma_j \frac{dy_j}{dt} = -\frac{\partial H}{\partial x_j}, \tag{2.1.21}$$

where $j = 1, \dots, N$ (there is no sum on j). Introduce the new variables

$$x_i' = \sqrt{|\Gamma_i|} x_i, \qquad y_i' = \sqrt{|\Gamma_i|} \operatorname{sgn}(\Gamma_i) y_i, \qquad i = 1, \dots, N,$$

where $\operatorname{sgn}(\Gamma_i)$ is 1 if $\Gamma_i > 0$, and is -1 otherwise. Then (2.1.19) is equivalent to the following system of Hamiltonian equations

$$\frac{dx_i'}{dt} = \frac{\partial H}{\partial y_i'}, \qquad \frac{dy_i'}{dt} = -\frac{\partial H}{\partial x_i'}, \qquad i = 1, \dots, N, \tag{2.1.22}$$

with Hamiltonian H and generalized coordinates (x_i', y_i'). As in elementary mechanics,

$$\frac{dH}{dt} = \sum_{i=1}^{N} \frac{\partial H}{\partial x_i'} \frac{dx_i'}{dt} + \sum_{i=1}^{N} \frac{\partial H}{\partial y_i'} \frac{dy_i'}{dt} = 0,$$

[2] See C. Anderson and C. Greengard, *On Vortex Methods, SIAM J. Sci. Statist. Comput.* **22** [1985], 413.

that is, H is a constant of the motion. A consequence of this fact is that if the vortices all have the same sign they cannot collide. If $\|\mathbf{x}_i - \mathbf{x}_j\| \neq 0, i \neq j$ at $t = 0$, then this remains so for all time because if $\|\mathbf{x}_i - \mathbf{x}_j\| \to 0, H$ becomes infinite.

This Hamiltonian system is of importance in understanding how vorticity evolves and organizes itself.[3]

Let us now generalize the situation and imagine our N vortices moving in a domain D with boundary ∂D. We can go through the same construction as before, but we have to modify the flow $\mathbf{u}_j$ of the jth vortex in such a way that $\mathbf{u} \cdot \mathbf{n} = 0$, that is, the boundary conditions appropriate to the Euler equations hold. We can arrange this by adding a potential flow $\mathbf{v}_j$ to $\mathbf{u}_j$ such that $\mathbf{u}_j \cdot \mathbf{n} = -\mathbf{v}_j \cdot \mathbf{n}$ on ∂D. In other words, we choose the stream function ψ_j for the jth vortex to solve

$$\Delta \psi_j = -\xi_j = -\Gamma_j \delta(\mathbf{x} - \mathbf{x}_j) \quad \text{with } \frac{\partial \psi_j}{\partial n} = 0 \text{ on } \partial D.$$

This is equivalent to requiring $\psi_j(\mathbf{x}) = -\Gamma_j G(\mathbf{x}, \mathbf{x}_j)$ where G is the Green's function for the Neumann problem for the region D. This procedure will appropriately modify the function $(1/2\pi) \log \|\mathbf{x} - \mathbf{x}_j\|$ and allow the analysis to go through as before.

For example, suppose D is the upper half-plane $y \geq 0$. Then we get G by the reflection principle:

$$G(\mathbf{x}, \mathbf{x}_j) = \frac{1}{2\pi} \left(\log \|\mathbf{x} - \mathbf{x}_j\| + \log \|\mathbf{x} - \hat{\mathbf{x}}_j\| \right),$$

where $\hat{\mathbf{x}}_j = (x_j, -y_j)$ is the reflection of $\mathbf{x}_j$ across the x-axis (see Figure 2.1.10). For the Neumann-Green's functions for other regions the reader may consult textbooks on partial differential equations.

Consider again Euler's equations in the form

$$\Delta \psi = -\xi, \quad u = \partial_y \psi, \quad v = -\partial_x \psi, \quad \frac{D\xi}{Dt} = 0.$$

One can write

$$\psi = -\int \xi(\mathbf{x}') G(\mathbf{x}, \mathbf{x}') \, d\mathbf{x}',$$

where $G(\mathbf{x}, \mathbf{x}') = \frac{1}{2}\pi \log \|\mathbf{x} - \mathbf{x}'\|$, and set $u = \partial_y \psi$, $v = -\partial_x \psi$. The resulting equations resemble the equations just derived for a system of point

[3]The Euler equations themselves form a Hamiltonian system (this is explained, along with references, in R. Abraham and J. E. Marsden, *Foundations of Mechanics,* 2nd Edition [1978]), and the Hamiltonian nature of the vortex approximation is consistent with this. See also J. E. Marsden and A. Weinstein, Coadjoint orbits, vortices and Clebsch variables for incompressible fluids, *Physica* **7D** [1983], 305–323.

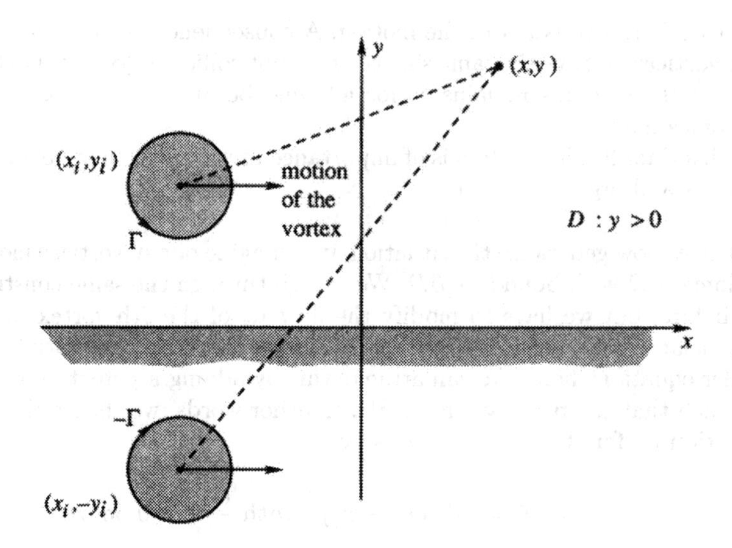

FIGURE 2.1.10. The stream function at (x, y) is the superposition of those due to vortices with opposite circulations located at (x_i, y_i) and $(x_i, -y_i)$.

vortices. The integral for ψ here resembles the formula for ψ in the point vortex system somewhat as an integral resembles one of its Riemann sum approximations. This suggests that an ideal incompressible flow can be approximated by the motion of a system of point vortices. There are in fact theorems along these lines.[4] Vortex systems provide both a useful heuristic tool in the analysis of the general properties of the solutions of Euler's equations, and a useful starting point for the construction of practical algorithms for solving these equations in specific situations.

One can ask if there is a similar construction in three dimensions. First of all, one can seek an analogue of the superposition of stream functions from point potential vortices. Given $\mathbf{u}$ satisfying $\operatorname{div} \mathbf{u} = 0$, there is a vector field $\mathbf{A}$ such that $\operatorname{div} \mathbf{A} = 0$ and such that $\mathbf{u} = \operatorname{curl} \mathbf{A}$, and therefore $\Delta \mathbf{A} = -\boldsymbol{\xi}$. In three dimensions, Green's function for the Laplacian is given by

$$G(\mathbf{x}, \mathbf{x}') = -\frac{1}{4\pi} \frac{1}{\|\mathbf{x} - \mathbf{x}'\|}, \qquad \mathbf{x} \neq \mathbf{x}'.$$

[4]The discrete vortex method is discussed in L. Onsager, *Nuovo Cimento* **6** (Suppl.) [1949], 229; A. J. Chorin, *J. Fluid Mech.* **57** [1973], 781; and A. J. Chorin, *SIAM J. Sci. Statist. Comput.* **1** [1980], 1. Convergence of solutions of the discrete vortex equations to solutions of Euler's equations as $N \to \infty$ is discussed in O. H. Hald, *SIAM J. Numer. Anal.* **16** [1979], 726; T. Beale and A. Majda, *Math. Comp.* **39** [1982], 1–28, 29–52; and K. Gustafson and J. Sethian, *Vortex Flows,* SIAM Publications, 1991.

Then we can represent $\mathbf{A}$ in terms of $\boldsymbol{\xi}$ by

$$\mathbf{A} = -\frac{1}{4\pi}\int \frac{\boldsymbol{\xi}(\mathbf{x}')}{s}\, dV(\mathbf{x}'),$$

where $s = \|\mathbf{x} - \mathbf{x}'\|$, and where $dV(\mathbf{x}')$ is the usual volume element in space. It is easy to check that $\mathbf{A}$ defined by the above integral satisfies the normalization condition $\operatorname{div} \mathbf{A} = 0$. Thus, because $\mathbf{u} = \operatorname{curl} \mathbf{A}$, we obtain

$$\mathbf{u}(\mathbf{x}) = \frac{1}{4\pi}\int \frac{\mathbf{s} \times \boldsymbol{\xi}'}{s^3}\, dV(\mathbf{x}'),$$

where $\mathbf{s} = \mathbf{x} - \mathbf{x}$ and $\boldsymbol{\xi}' = \boldsymbol{\xi}(\mathbf{x}')$. Suppose that we have a vortex line C in space with circulation Γ (see Figure 2.1.11) and we assume that the vorticity field $\boldsymbol{\xi}$ is concentrated on C only, that is, the flow is potential outside the filament C. Then $\mathbf{u}(\mathbf{x})$ can be written as

$$\mathbf{u}(\mathbf{x}) = \frac{1}{4\pi}\int_C \frac{\mathbf{s} \times \Gamma\, d\mathbf{s}}{s^3}$$

where $d\mathbf{s}$ is the line element on C.

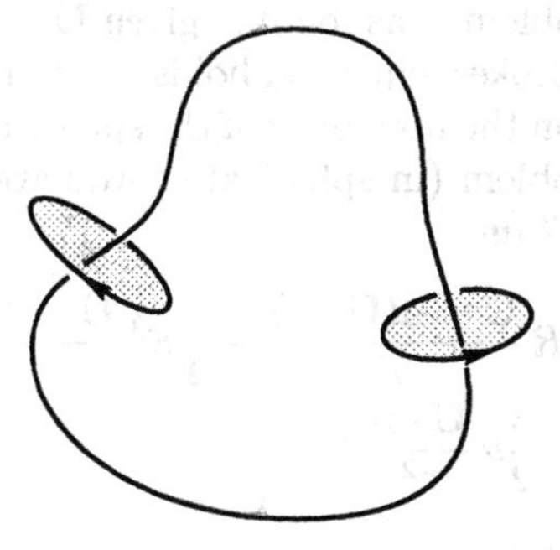

FIGURE 2.1.11. The flow induced by a vortex filament.

Exercises

◇ **Exercise 2.1-1** If $f : D \to D'$ is a conformal transformation (an analytic function that is one to one and onto), it can be used to transform one complex potential to another. Use $f(z) = z + a^2/z$ (which takes the exterior of the disc of radius a in the upper half-plane to the upper half-plane) and the complex potential in the upper half-plane to generate formula (2.1.11).

◇ **Exercise 2.1-2** Let $F(z) = z^2$ be a complex potential in the first quadrant. Sketch some streamlines and the curves ϕ = constant, ψ = constant, where $F = \phi + i\psi$. What is the force exerted on the walls?

◇ **Exercise 2.1-3** Use conformal maps to find a formula for potential flow over the plate in Figure 2.1.12. What is the force exerted on this plate?

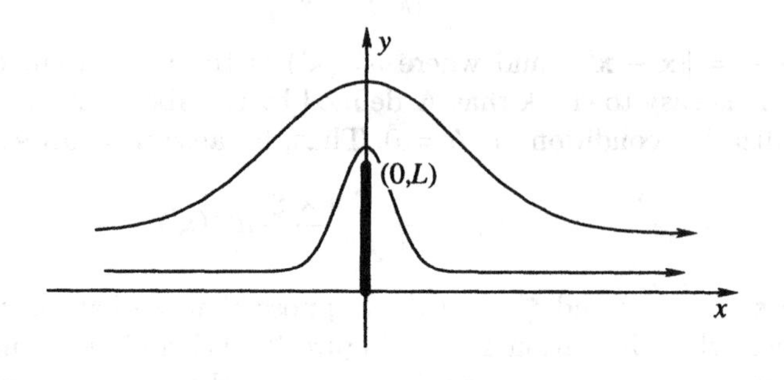

FIGURE 2.1.12. Flow over a vertical plate.

◇ **Exercise 2.1-4** Let a spherical object move through a fluid in $\mathbb{R}^3$. For slow velocities, assume Stokes' equations apply. Take the point of view that the object is stationary and the fluid streams by. The setup for the boundary value problem is as follows: given $\mathbf{U} = (U, 0, 0), U$ constant, find $\mathbf{u}$ and p such that Stokes' equation holds in the region exterior to a sphere of radius R, $\mathbf{u} = \mathbf{0}$ on the boundary of the sphere and $\mathbf{u} = \mathbf{U}$ at infinity. The solution to this problem (in spherical coordinates centered in the object) is called ***Stokes' Flow***:

$$\begin{aligned} \mathbf{u} &= -\frac{3}{4}R\,\frac{\mathbf{U} + \mathbf{n}(\mathbf{U}\cdot\mathbf{n})}{r} - \frac{1}{4}R^3\,\frac{\mathbf{U} - 3\mathbf{n}(\mathbf{U}\cdot\mathbf{n})}{r^3} + \mathbf{U}, \\ p &= p_0 - \frac{3}{2}\nu\,\frac{\mathbf{U}\cdot\mathbf{n}}{r^2}\,R, \end{aligned} \tag{2.1.23}$$

where p_0 is constant and $\mathbf{n} = \mathbf{r}/r$.

(a) Verify this solution.

(b) Show that the drag is $6\pi R\nu U$ and there is no lift.

(c) Show there is net outflow at infinity (an infinite wake).

◇ **Exercise 2.1-5** Because of the difficulties in Exercise 2.1-4, Oseen in 1910 suggested that Stokes' equations be replaced by

$$-\nu\Delta\mathbf{u} + (\mathbf{U}\cdot\nabla)\mathbf{u} = -\frac{1}{\rho}\,\operatorname{grad} p,$$

with $\operatorname{div}\mathbf{u} = 0$, where $\mathbf{u}$ represents the true velocity minus $\mathbf{U}$. This amounts to linearizing the Navier–Stokes equations about $\mathbf{U}$, whereas Stokes' equations may be viewed as a linearization about $\mathbf{0}$. One would thus conjecture that Oseen's equations are good where the flow is close to the free stream velocity $\mathbf{U}$ (away from the sphere) and that Stokes' equations are good where the velocity is $\mathbf{0}$ (near the sphere). The solution of Oseen's equations in the region exterior to a sphere in $\mathbb{R}^3$ can be found in Lamb's book. Show that drag on the sphere for the Oseen solution is $F = 6\pi RU\nu(1+\frac{3}{8}R)$, where $R = UR/\nu$ is the Reynolds number. Thus, there is a difference of the order R in the Stokes and Oseen drag forces.

Notes on Exercise 2.1-4 and Exercise 2.1-5: If D is bounded with smooth boundary, then there exists at most one solution to Stokes' equations. See Ladyzhenskaya's book listed in the Preface. In the exterior of a bounded region in $\mathbb{R}^3$ there exists a unique solution to Stokes' equations. The situation in $\mathbb{R}^2$ is different; in fact, we have the following strange situation:

Stokes' Paradox *There is no solution to Stokes' equations in $\mathbb{R}^2$ in the region exterior to a disc (with reasonable boundary conditions).*[5]

Stokes' paradox does not apply to the Oseen or Navier–Stokes equations in $\mathbb{R}^2$ or $\mathbb{R}^3$. However, Filon in 1927 pointed out that for other reasons Oseen's equations also lead to unacceptable results. The example he gives is a skewed ellipse in a free stream. Computation of the moment exerted on the ellipse reveals that it is infinite! This is not so surprising in view of the fact that Oseen's equations represent linearization about the free stream. One cannot expect them to give good results around the obstacle because the equations contain errors there of order U^2.

2.2 Boundary Layers

Consider the Navier–Stokes equations

$$\left.\begin{aligned} \partial_t \mathbf{u} + (\mathbf{u}\cdot\nabla)\mathbf{u} &= -\operatorname{grad} p + \frac{1}{R}\,\Delta\,\mathbf{u}, \\ \operatorname{div}\mathbf{u} &= 0, \\ \mathbf{u} &= \mathbf{0} \quad \text{on } \partial D, \end{aligned}\right\} \tag{2.2.1}$$

and assume the Reynolds number R is large. We ask how different a flow governed by (2.2.1) is from one governed by the Euler equations for incompressible ideal flow:

$$\left.\begin{aligned} \partial_t \mathbf{u} + (\mathbf{u}\cdot\nabla)\mathbf{u} &= -\operatorname{grad} p, \\ \operatorname{div}\mathbf{u} &= 0, \\ \mathbf{u}\cdot\mathbf{n} &= 0 \quad \text{on } \partial D. \end{aligned}\right\} \tag{2.2.2}$$

---

[5]See Birkhoff's book, and J. Heywood, *Arch. Rational Mech. Anal.* **37** [1970], 48–60, and *Acta Math.* **129** [1972], 11–34.

Imagine that both flows coincide at $t = 0$ and, say, are irrotational, that is, $\boldsymbol{\xi} = \mathbf{0}$. Thus, under (2.2.2) the flow stays irrotational, and thus is a potential flow. However, we claim that the presence of the (small) viscosity term $(1/R)\,\Delta\,\mathbf{u}$ and the difference in the boundary conditions have the following effects:

1. The flow governed by (2.2.2) is drastically modified near the wall in a region with thickness proportional to $1/\sqrt{R}$.

2. The region in which the flow is modified may separate from the boundary.

3. This separation will act as a source of vorticity.

Whereas the region referred to in 1 gets arbitrarily small as R increases, its effects via 2 and 3 may persist and not diminish in the limit $R \to \infty$. A model for the first effect is given in the following:

Example 1 Consider the equation

$$\frac{dy}{dx} = a, \qquad y(1) = 1, \tag{2.2.3}$$

where a is a constant and x ranges between $x = 0$ and $x = 1$. The solution of (2.2.3) is clearly

$$y = a(x - 1) + 1.$$

Now add a "viscosity" term and a boundary condition to (2.2.3) as follows:

$$\epsilon\frac{d^2y}{dx^2} + \frac{dy}{dx} = a, \quad \text{with } y(0) = 0, \;\; y(1) = 1. \tag{2.2.4}$$

As with the Navier–Stokes and Euler equations, (2.2.4) differs from (2.2.3) by the addition of a small constant times a higher-order term, as well as by the necessary addition of an extra boundary condition. The solution of (2.2.4) is verified to be

$$y = \frac{1-a}{1-e^{-1/\epsilon}}\left(1 - e^{-x/\epsilon}\right) + ax.$$

For $0 < a < 1$, y is graphed in Figure 2.2.1.
For ϵ small and $x > \epsilon$, the two solutions are close together. The region where they are drastically different is confined to the interval $[0, \epsilon]$, which we can refer to as the ***boundary layer***. Note that as $\epsilon \to 0$, the boundary layer shrinks to zero, but the maximum difference between the solutions remains constant. ♦

Next we consider a special case in which (2.2.1) and (2.2.2) can be solved and the boundary layer can be seen explicitly.

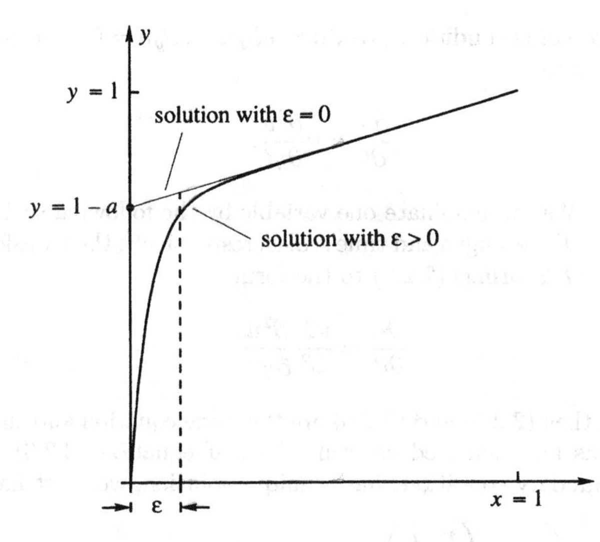

FIGURE 2.2.1. Comparing solutions of (2.2.3) and (2.2.4).

Example 2 Consider two-dimensional flow in the upper half-plane $y \geq 0$, and suppose the boundary $y = 0$ ("the plate") is rigid and the velocity at $y = \infty$ is parallel to the x-axis and has magnitude U. See Figure 2.2.2.

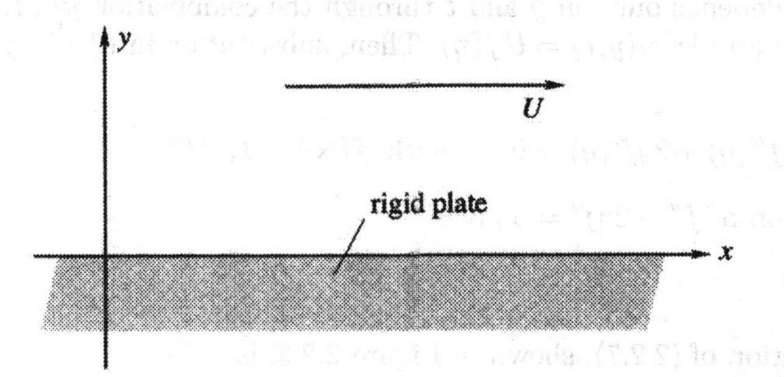

FIGURE 2.2.2. Flow over a flat plate.

We seek a flow satisfying (2.2.1) that is parallel to the plate and independent of x, that is, $\mathbf{u}$ has the form $\mathbf{u} = (u(y,t), 0)$, and, with constant pressure, so $\operatorname{grad} p = 0$. The appropriate solution of Euler's equations (2.2.2) is the solution $\mathbf{u}(x, y, t) = (U, 0)$.

The boundary conditions for (2.2.1) are

$$u(0,t) = 0 \quad \text{and} \quad u(\infty, t) = U.$$

Under the preceding conditions, $\mathbf{u} \cdot \nabla \mathbf{u} = u\partial_x \mathbf{u} + v\partial_y \mathbf{u} = 0$, and so equation (2.2.1) reduces to

$$\frac{\partial u}{\partial t} = \nu \frac{\partial^2 u}{\partial y^2}, \tag{2.2.5}$$

where $\nu = \frac{1}{R}$. We can eliminate one variable by the following scaling argument: If L and T are length and time scales, respectively, the transformation $y' = y/L$, $t' = t/T$ brings (2.2.5) to the form

$$\frac{\partial u}{\partial t'} = \frac{\nu T}{L^2} \frac{\partial^2 u}{\partial y'^2}. \tag{2.2.6}$$

If $L^2/T = 1$, then (2.2.5) and (2.2.6) are the same equation and the boundary conditions are unaltered as well. Thus, if equation (2.2.6) with the preceding boundary conditions has a unique solution, we must have

$$u\left(\frac{y}{L}, \frac{t}{T}\right) = u(y,t) \quad \text{if } L^2 = t.$$

Picking $T = t$ and $L = \sqrt{t}$, we obtain

$$u\left(\frac{y}{\sqrt{t}}, 1\right) = u(y,t).$$

Thus, u depends only on y and t through the combination $y/\sqrt{t}$. Set $\eta = y/(2\sqrt{\nu t})$ and let $u(y,t) = Uf(\eta)$. Then, substitution in (2.2.5) yields the equation

$$f''(\eta) + 2\eta f'(\eta) = 0 \qquad \text{with } f(\infty) = 1,\ f(0) = 0. \tag{2.2.7}$$

Integration of $f'' + 2\eta f' = 0$ gives

$$f'(\eta) = ce^{-\eta^2}, \qquad c \text{ a constant.}$$

The solution of (2.2.7), shown in Figure 2.2.3, is

$$u = U \operatorname{erf}(\eta),$$

where the ***error function*** is defined by

$$\operatorname{erf}(\eta) = \frac{2}{\sqrt{\pi}} \int_0^\eta e^{-s^2}\, ds.$$

In matching the boundary condition at infinity, we have used the following basic identity for the error function:

$$\int_0^\infty e^{-s^2}\, ds = \frac{\sqrt{\pi}}{2}.$$

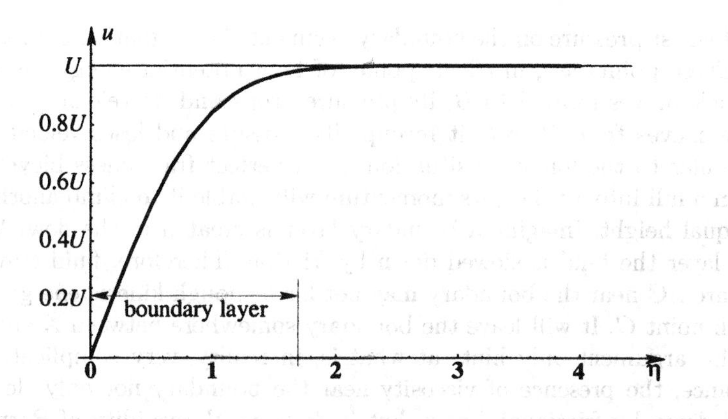

FIGURE 2.2.3. The boundary layer has thickness proportional to $(t/R)^{1/2}$.

The graph of u as a function of y for fixed $t > 0$ is obtained by rescaling the above graph. The region in which u departs significantly from the constant Euler flow is again called the ***boundary layer*** and is illustrated in the figure. By the preceding scaling argument, this thickness is proportional to $\sqrt{\nu t} = \sqrt{t}/\sqrt{R}$. Thus, for fixed time, the boundary layer decreases as $1/\sqrt{R}$, which was our first contention. ♦

Our second contention was that the modified flow in the boundary layer may separate from the body.

We can give a heuristic argument justifying this contention by considering potential flow around a cylinder (Figure 2.2.4). By Bernoulli's theorem,

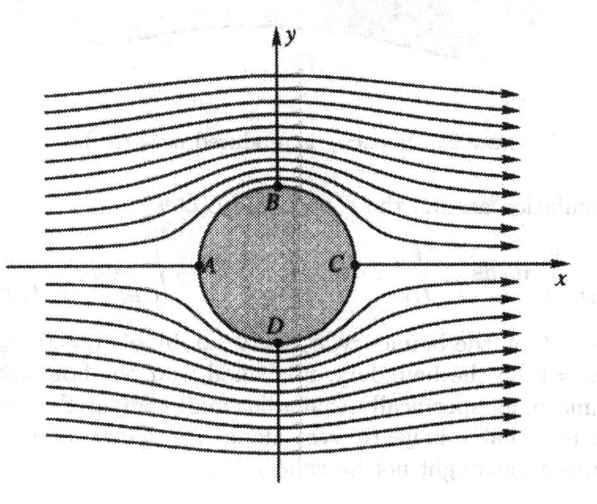

FIGURE 2.2.4. Illustration for the discussion of separation.

the highest pressure on the boundary occurs at the stagnation points A and C. These points are, in effect, points of high potential energy. As a fluid particle moves from A to B, its pressure drops and its velocity increases. As it moves from B to C it recoups its pressure and loses velocity. This is similar to the following situation: As a perfect frictionless bicycle goes down a hill into a valley, its momentum will enable it to climb another hill of equal height. Imagine a boundary layer is created in the flow. Within this layer the fluid is slowed down by friction. Therefore, fluid traversing the arc AC near the boundary may not have enough kinetic energy left to reach point C. It will leave the boundary somewhere between A and C

This argument only hints at what is, in reality, very complicated. For instance, the presence of viscosity near the boundary not only slows the fluid down by frictional forces, but it destroys the validity of Bernoulli's theorem! Nevertheless, the preceding argument probably does have some merit; in fact, experiments show that flows really can undergo boundary layer separation.

Our third contention was that the boundary layer produces vorticity. To see this, imagine flow near the boundary, as shown in Figure 2.2.5.

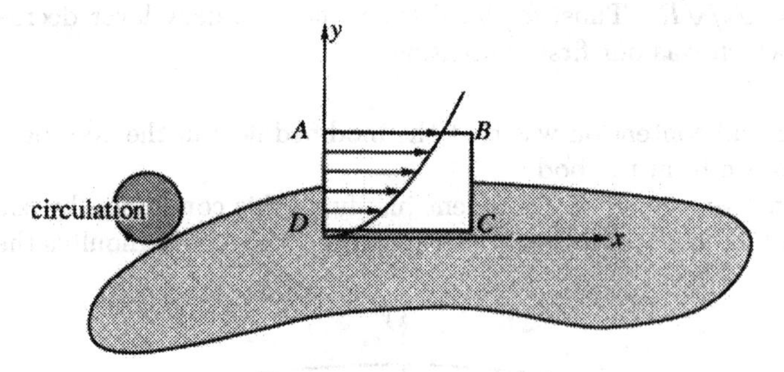

FIGURE 2.2.5. Vorticity is produced near the boundary.

The circulation around the contour $ABCD$ is

$$\int_{ABCD} \mathbf{u} \cdot ds = \int_{DA} v\, dy + \int_{AB} u\, dx - \int_{CB} v\, dy - \int_{DC} u\, dx.$$

Because $u = 0$ on the boundary, $\partial_x u = 0$; so, by $\operatorname{div} \mathbf{u} = 0$, $\partial_y v = 0$. Thus, because $v = 0$ on the boundary, v is small near the boundary. Let us, in fact, assume more specifically that v is small compared to the value of u along AB and that u is nearly zero along DC. (Near points of separation these assumptions might not be valid.) Thus,

$$\int_{ABCD} \mathbf{u} \cdot ds \cong \int_{AB} u\, dx > 0$$

and so the flow has circulation and therefore vorticity.

Next we shall make some approximations in the Navier–Stokes equations that seem to be intuitively reasonable near the boundary, away from points of separation. This will yield the Prandtl boundary layer equations.

We now consider two-dimensional incompressible homogeneous flow in the upper half-plane $y > 0$. We write the Navier–Stokes equations as

$$\left.\begin{aligned} \partial_t u + u\,\partial_x u + v\,\partial_y u &= -\partial_x p + \frac{1}{R}\Delta u, \\ \partial_t v + u\,\partial_x v + v\,\partial_y v &= -\partial_y p + \frac{1}{R}\Delta v, \\ \partial_x u + \partial_y v &= 0, \\ u = v &= 0 \quad \text{on } \partial D. \end{aligned}\right\} \tag{2.2.8}$$

Let us assume that the Reynolds number R is large and that a boundary layer of thickness δ develops. From our previous discussions, we expect that $\delta \sim 1/\sqrt{R}$, where $\sim$ means "of the same order as." See Figure 2.2.6.

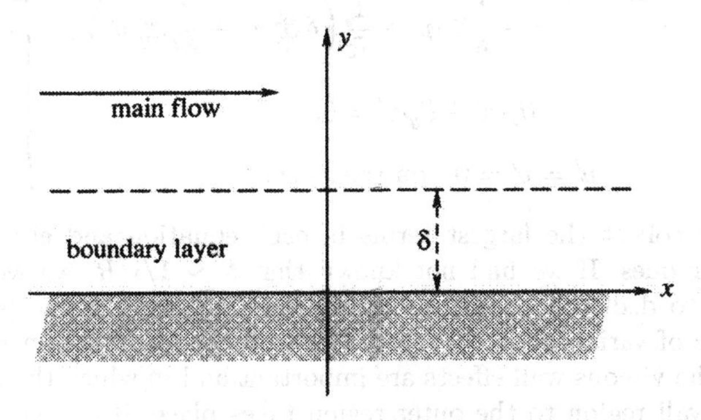

FIGURE 2.2.6. The boundary layer thickness δ is of order $1/R^{1/2}$.

We also expect, as discussed earlier, v to be small in the boundary layer. We want to perform a change of variables that will focus attention on the thin layer near the wall in which viscous effects are important. The layer is thin, of order δ. Write $y' = y/\delta$, so that when y varies between 0 and δ, y' varies between 0 and 1. We expect that u is of order 1 at a distance of order δ from the wall, because at the edge of the layer, the flow should be effectively inviscid by definition (we are focusing on the region where the viscous effects are important). We claim that v is small at a distance δ from the wall. Indeed, at the wall, $v = 0$ and so inside the fluid,

$$v(x, y) \approx v(x, 0) + y\frac{\partial v}{\partial y} = y\frac{\partial v}{\partial y};$$

y is of order δ at most, and $\partial v/\partial y$ is of order 1 because

$$\frac{\partial v}{\partial y} + \frac{\partial u}{\partial x} = 0$$

and

$$\frac{\partial u}{\partial x} \text{ is of order 1.}$$

Thus, we perform the change of variables

$$x' = x, \quad y' = \frac{y}{\delta}, \quad t' = t, \quad u' = u, \quad v' = \frac{v}{\delta}, \quad p' = p.$$

Equations (2.2.8) become

$$\left.\begin{aligned}
&\partial_{t'} u' + u'\,\partial_{x'} u' + v'\,\partial_{y'} u' \\
&\qquad = -\partial_{x'} p' + \frac{1}{R}\left(\partial_{x'}^2 u' + \frac{1}{\delta^2}\partial_{y'}^2 u'\right), \\
&\delta\,\partial_{t'} v' + \delta u'\,\partial_{x'} v' + \delta v'\,\partial_{y'} v' \\
&\qquad = -\frac{1}{\delta}\partial_{y'} p' + \frac{1}{R}\left(\delta\,\partial_{x'}^2 v' + \frac{\delta}{\delta^2}\partial_{y'}^2 v'\right), \\
&\qquad \partial_{x'} u' + \partial_{y'} v' = 0, \\
&\quad u' = v' = 0 \quad \text{on the } x\text{-axis.}
\end{aligned}\right\} \tag{2.2.9}$$

We now collect the largest terms in each equation and eliminate the lower-order ones. If we had not known that $\delta \sim 1/\sqrt{R}$, we would have been able to deduce it from the scaled equation. Indeed, the purpose of the change of variables $y' = y/\delta$, etc., is to focus attention on the region in which the viscous wall effects are important and in which the transition from the wall region to the outer region takes place. If δ is smaller than $O(1/\sqrt{R})$, then only the viscous effects survive as the dominant terms in our equation $(2.2.9)_1$, and so we are looking too close to the wall. If δ is larger than $O(1/\sqrt{R})$ the viscous effects disappear altogether and we are looking at a region that is too large. Thus, $\delta = O(1/\sqrt{R})$ is the only choice. From these arguments, it is plausible to believe that in the boundary layer, a good approximation to the Navier–Stokes equations is achieved by the following equations:

$$\left.\begin{aligned}
\partial_t u + u\,\partial_x u + v\,\partial_y u &= -\partial_x p + \frac{1}{R}\partial_y^2 u, \\
\partial_y p &= 0, \\
\partial_x u + \partial_y v &= 0, \\
u = 0 = v &\quad \text{on the } x\text{-axis.}
\end{aligned}\right\} \tag{2.2.10}$$

These are called the ***Prandtl boundary layer equations.***

One usually hopes, or assumes, that the following is true: If $\mathbf{u}(x, y, t)$ is a solution of the Navier–Stokes equations and $\mathbf{u}_p(x, y, t)$ is the corresponding solution of the boundary layer equations with the same initial conditions, then for some constant $\alpha > 0$ and some constant C,

$$\|\mathbf{u}(x, y, t) - \mathbf{u}_p(x, y, t)\| \leq \frac{C}{R^\alpha}$$

for $0 \leq y \leq \delta$ as $R \to \infty$. Here $\|\cdot\|$ is a norm (to be found) on the space of velocity fields. The boundary layer equations have been quite successful and so one expects some estimate like this to be valid, at least under some reasonable conditions. While some partial results verify this,[6] a full mathematical proof under satisfactory hypotheses is not available.

Let us derive a few consequences of the boundary layer equations (2.2.10). First of all, $(2.2.10)_2$ implies that p is a function of x alone. This implies that if the pressure has been determined outside the boundary layer, $(2.2.10)_2$ determines it inside the boundary layer.

Secondly, let us derive an equation for the propagation of vorticity in the boundary layer. From the derivation of the equations, we see that $\partial_x v$ is to be neglected in comparison to $\partial_y u$, and thus we have

$$\xi = \partial_x v - \partial_y u = -\partial_y u.$$

Therefore, from $(2.2.10)_1$,

$$\begin{aligned}
\partial_t \xi = -\partial_y \, \partial_t u &= -\partial_y \left(-\partial_x p + \frac{1}{R} \partial_y^2 u - u \, \partial_x u - v \, \partial_y u \right) \\
&= \frac{1}{R} \partial_y^2 \xi - v \, \partial_y \xi - u \, \partial_x \xi + (\partial_x u + \partial_y v) \partial_y u \\
&= \frac{1}{R} \partial_y^2 \xi - v \, \partial_y \xi - u \, \partial_x \xi,
\end{aligned}$$

because $\partial_x u + \partial_y v = 0$. Thus,

$$\frac{D\xi}{Dt} = \frac{1}{R} \partial_y^2 \xi \qquad (2.2.11)$$

describes the propagation of vorticity in the boundary layer. We can interpret (2.2.11) by saying that the vorticity is convected downstream and, at

---

[6] See O. A. Oleinik, *Soviet Math. Dokl.* [1968], for existence and uniqueness theorems. P. C. Fife *(Arch. Rational Mech. Anal.* **28** [1968], 184) has proved the above estimate for $\alpha = 1/2$ assuming a forward pressure gradient, steady flow, and some other technical conditions. The boundary layer equations for curved boundaries are more complicated (see below and S. L. Goldstein, *Modern Developments in Fluid Mechanics* (two volumes), Dover [1965]). To the authors' knowledge, no theorems of existence or approximation have been proved in this case.

the same time, is diffused vertically into the fluid. For the full Navier–Stokes equations in two dimensions, recall that

$$\frac{D\xi}{Dt} = \frac{1}{R}\Delta\xi$$

(see equation (1.3.11) of Chapter **1**). In the boundary layer approximation (2.2.11) only the y part of the Laplacian has survived. Notice also that in the full equations we recover $\mathbf{u}$ from ξ by the equations

$$\Delta\psi = -\xi \quad \text{with} \quad \psi = 0 \quad \text{on } \partial D,$$

and

$$u = \partial_y\psi, \qquad v = -\partial_x\psi$$

(see equation (1.3.12) of Chapter **1**). In the boundary layer approximation the situation is much simpler:

$$\xi = -\partial_y u, \quad \text{so} \quad u = -\int_0^y \xi\, dy.$$

Notice that the flow past the flat plate described at the beginning of this section is not only a solution of the Navier–Stokes equations but is a solution of the Prandtl equations as well. That solution was

$$u(y,t) = U\text{erf}\left(\frac{y}{2\sqrt{\nu t}}\right).$$

The vorticity for this flow $\xi = -\partial_y u$ is shown in Figure 2.2.7. The vorticity is everywhere negative corresponding to clockwise circulation, which agrees with our intuition.

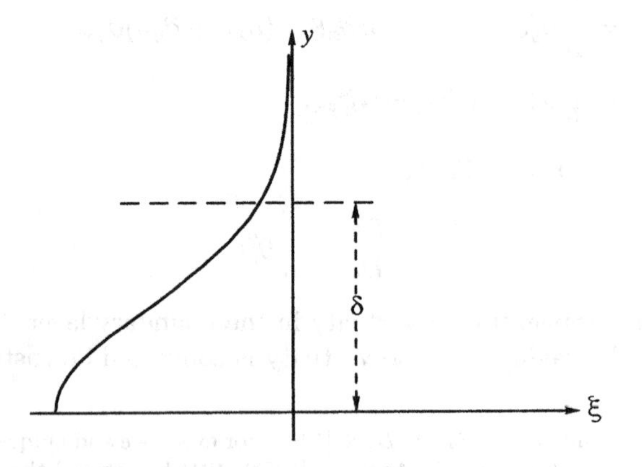

FIGURE 2.2.7. Vorticity in the boundary layer.

Our derivation of the boundary layer equations assumed the wall was flat. For curved walls one can imagine using coordinate systems with $y = 0$ on the wall and $x =$ constant giving the normal to the wall. One can show (see the Goldstein reference in the preceding footnote) that $(2.2.10)_1$, $(2.2.10)_3$ and $(2.2.10)_4$ are unaltered, but $(2.2.10)_2$ should be modified to

$$ku^2 = \partial_y p, \qquad (2.2.10)'_2$$

where k is the curvature of the wall. See Figure 2.2.8.

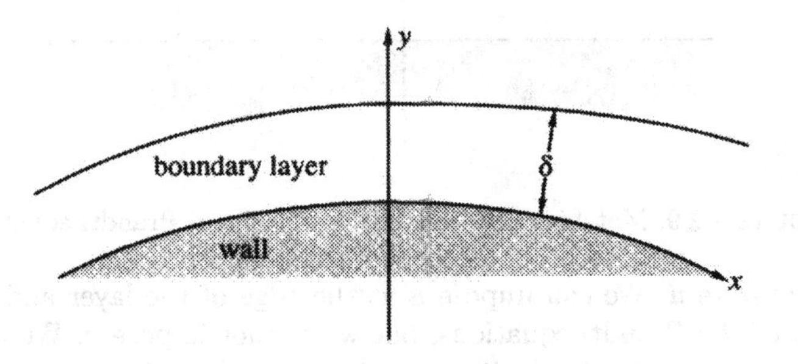

FIGURE 2.2.8. Boundary layer for a curved wall.

This means that p is no longer a function of x alone. Of course, for δ small (R large) and fixed curvature of the wall, the pressure change across the boundary layer is negligible. However, $(2.2.10)'_2$ may still be significant in affecting the dependence of p on x.

We have not yet discussed what boundary conditions should (or could) be imposed on the solution of the Prandtl equations beyond those at the wall, $u = v = 0$; these boundary conditions are essential for the derivation of the equations to make sense. We shall see later that far from the body, we may impose u but not v. Boundary conditions may also be required on the left and on the right. For a discussion, see the next section.

Now we want to discuss how a knowledge of boundary layer theory might be useful in constructing solutions of the Navier–Stokes equations. The overall method is to use the Prandtl equations in the boundary layer, the inviscid equations outside the boundary layer, and to try to match these two solutions to produce an approximate solution to the Navier–Stokes equations. To carry out such a program, there are several alternative procedures, four of which are listed here.

Strategy 1 Match the actual velocity field $\mathbf{u}$ of the two solutions at a distance $\delta \sim 1/\sqrt{R}$ from the wall (Figure 2.2.9).

Difficulties On the line $y = \delta$ the flows may not be parallel, so matching $\mathbf{u}$ or some other flow quantity may not be easy.

Strategy 2 Require that the Euler solution outside the layer and the Prandtl solution match as $R \to \infty$. As $R \to \infty$, the layer becomes thinner; we can impose v at the boundary for the Euler equations, but we

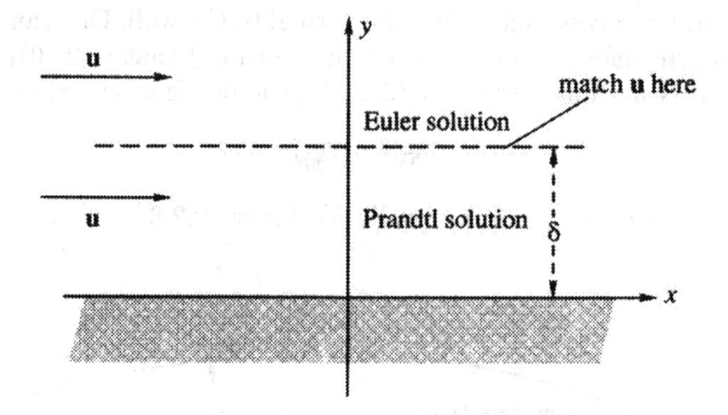

FIGURE 2.2.9. Matching solutions of the Euler and Prandtl equations.

cannot impose u. We can impose u at the edge of the layer and obtain a solution of the Prandtl equations, but we cannot impose v. We may wish to identify the vertical velocity v at the edge of the layer obtained from the Prandtl equations with the boundary condition required for the Euler equations, and conversely, use the tangential velocity v obtained from the Euler equations as the velocity at the edge of the layer imposed on the Prandtl equations.

Difficulties The calculations are difficult, and the results are insufficiently informative when separation occurs (see below).

Strategy 3 We can match the inner (boundary layer) and outer (Euler) solutions over a region whose height is of order $1/R^a$, $0 < a < 1/2$ (Figure 2.2.10). The matching can be done by some smooth transition between the Euler and Prandtl solutions.

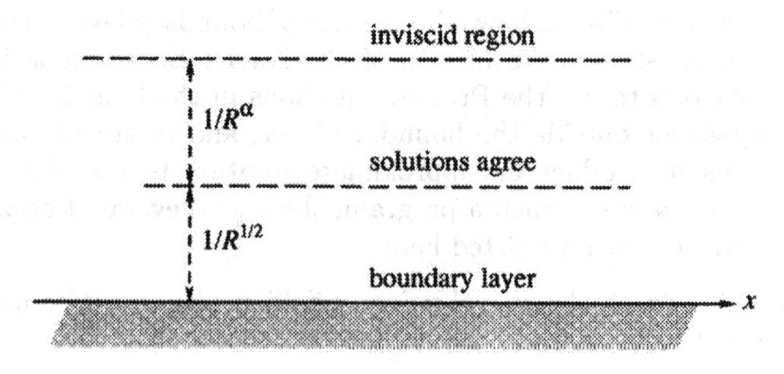

FIGURE 2.2.10. Matching method in strategy 3.

Difficulties The calculations are enormously complicated and require choices (of a for example) for which a rationale is not always readily available.

Strategy 4 As both the Euler and Prandtl equations are solved and matched by some method, check (by means of a computer) whether or not the assumptions underlying the derivation of the Prandtl equations are valid. If not, modify the equations.

Difficulties Same matching problems as before, but also modified boundary layer equations might be needed.

Techniques such as these are in fact fairly extensively developed in the literature.[7] The scope of the difficulties involved, both theoretical and practical, should be evident.

It is observed experimentally that when a boundary layer develops near the surface of an obstacle, the streamlines may break away from the boundary. This phenomenon is called ***boundary layer separation***. Figure 2.2.11 shows a typical velocity profile near the point of separation. Contemplation of this velocity field shows that it is plausible to identify the point of separation C with points where the circulation vanishes:

$$\left.\frac{\partial u}{\partial n}\right|_C = \xi = 0.$$

There are, however, no known theorems that apply to this situation that guarantee such points must correspond to separation points.

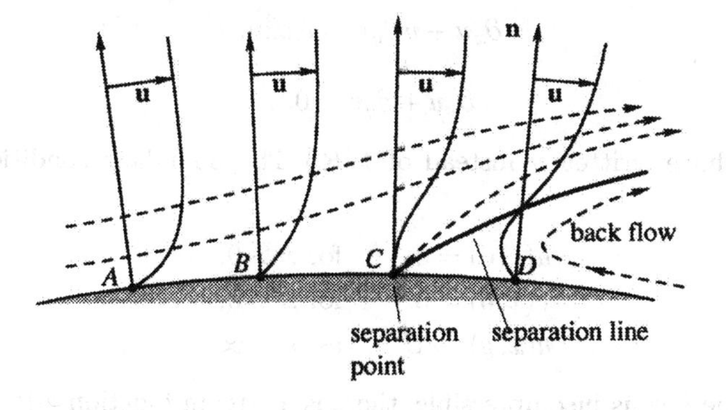

FIGURE 2.2.11. Separation of the boundary layer.

The phenomenon of separation can be used to justify our model of almost potential flow introduced in the previous section (see Figure 2.1.8). The two curves of vorticity springing from the boundary are imagined to arise from boundary layer separation. One would hope that boundary layer theory would predict the strength of these vortex lines and their points

[7]See, for instance, M. Van Dyke, *Perturbation Methods in Fluid Mechanics,* Parabolic Press.

of separation. Near the boundary, one should also be able to compute the pressure and hence the ***form drag***

$$-\int_{\text{boundary}} p\mathbf{n}\, ds.$$

This is the drag due to actual pressure forces on the obstacle. There will also be a ***skin drag*** due to frictional forces in the boundary layer. These two give the total drag. If one computes the drag from a nearly potential approximation, the skin drag turns out to be zero, because Euler's equations do not take into account the viscous forces. For large R, the skin drag is very often negligible compared to the form drag.

Example This example concerns *steady boundary layer flow on a flat plate of infinite width.* Consider a flat plate on the xz plane with its leading edge coinciding with $x = 0$, as shown in Figure 2.2.12. Let the y-axis be normal to the plate so that the fluid lies in the domain given by $y > 0$ and the flow is two dimensional. We then let the plate be washed by a steady uniform flow of velocity U in the positive x-direction. We assume that a steady boundary layer is developed after a short distance off the leading edge. Because the pressure is uniform outside the boundary layer, it is reasonable to seek a solution satisfying $\partial_x p = 0$ in the boundary layer. Because $\partial_x p = 0$, p is a constant everywhere. Hence, the steady boundary layer equations are

$$\begin{aligned} u\,\partial_x u + v\,\partial_y u &= \nu\,\partial_y^2 u, \\ \partial_x u + \partial_y v &= 0. \end{aligned} \tag{2.2.12}$$

(Here we have written ν instead of $1/R$.) The boundary conditions are given by

$$\begin{aligned} u(x,0) &= 0 \qquad \text{for } x > 0, \\ v(x,0) &= 0 \qquad \text{for } x > 0, \\ u(x,y) &\to U \qquad \text{as } y \to \infty. \end{aligned}$$

Because the flow is incompressible, there is a stream function related to $\mathbf{u}$ by the usual equations

$$u = \partial_y \psi, \qquad v = -\partial_x \psi.$$

We assume ψ to be of the form

$$\psi(x,y) = \sqrt{\nu U x}\; f(\eta),$$

where

$$\eta = y\sqrt{\frac{U}{\nu x}}.$$

This form may be derived through the use of a scaling argument similar to the one used earlier in §**2.1**. Thus,

$$u = Uf'(\eta), \quad v = \frac{1}{2}\sqrt{\frac{U\nu}{x}}(\eta f'(\eta) - f(\eta)). \qquad (2.2.13)$$

Substituting u, v from (2.2.13) into (2.2.12) and using the boundary conditions, we find the equations

$$ff'' + 2f''' = 0 \qquad (2.2.14)$$

with boundary conditions

$$f(0) = f'(0) = 0, \qquad f'(\infty) = 1.$$

Equation (2.2.14) has a numerically computable solution. In the fluid solution given by (2.2.13), notice that the velocity u depends only on η. The level curves of u are exactly the lines where η = constant. But level curves of η are of the form y^2 = (constant)x, that is, parabolas, as shown in Figure 2.2.12. ♦

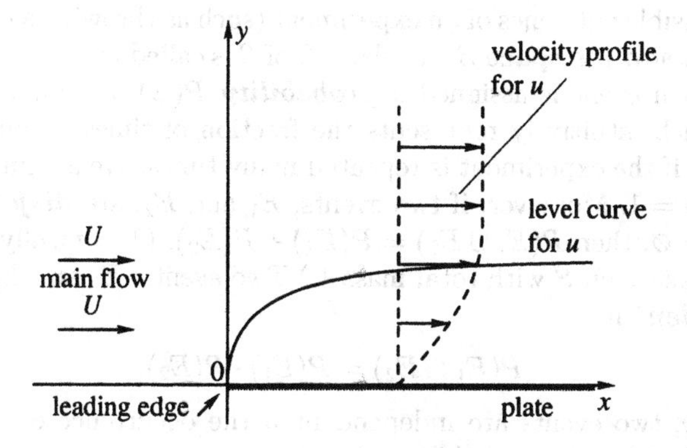

FIGURE 2.2.12. Boundary layer flow over a flat plate.

Exercises

◇ **Exercise 2.2-1** Show that the solution of

$$\epsilon y'' - y = 0$$

in the interval $(-1, 1)$ with the boundary conditions

$$y(-1) = 0, \quad y(1) = 1$$

is

$$y(x) = \frac{e^{(1+x)/\sqrt{\epsilon}} - e^{-(1+x)/\sqrt{\epsilon}}}{e^{2/\sqrt{\epsilon}} - e^{-2/\sqrt{\epsilon}}}.$$

Plot the graph of $y(x)$ for $\epsilon = 0.1, 0.01$ and 0.001.

◇ **Exercise 2.2-2** Derive the solution in Exercise 2.2-1 from that for (2.2.4) in the text by a scaling argument.

◇ **Exercise 2.2-3** Plot the shape of the graph of the function $f(\eta)$ that solves (2.2.14) with $f(0) = f'(0) = 0, f'(\infty) = 1$.

2.3 Vortex Sheets

This section describes a simple version of an algorithm for modifying a flow described by the Euler equations near the boundary to simulate the effects of boundary layers and thereby obtain an approximation to the Navier–Stokes equations. The method utilizes some concepts from probability theory, so we begin by summarizing some relevant facts, mostly without proofs.[8]

The possible outcomes of an experiment (such as throwing a die) form the points in a ***sample space*** S. A subset E of S is called an ***event***. We assume that to each event is assigned a ***probability*** $P(E)$, a number between 0 and 1 which intuitively represents the fraction of times an outcome in E will occur if the experiment is repeated many times. We assume, therefore, that $P(S) = 1$. Moreover, if two events, E_1 and E_2, are ***disjoint,*** that is, $E_1 \cap E_2 = \varnothing$, then $P(E_1 \cup E_2) = P(E_1) + P(E_2)$. (Technically, we assume P is a measure on S with total mass 1.) Two events, E_1 and E_2, are called ***independent*** if

$$P(E_1 \cap E_2) = P(E_1) \cdot P(E_2).$$

Intuitively, two events are independent if the occurrence of one of them has no effect on the probability of the occurrence of the other one. (For instance, in the toss of two dice marked #1 and #2, the events "a two on #1" and "a three or a four on #2" are independent.)

A ***random variable*** is a mapping

$$\eta : S \to \mathbb{R},$$

interpreted as a number attached to the outcome of an experiment. The ***expectation*** or ***mean*** of η is defined by

$$E(\eta) = \int_S \eta \, dP.$$

[8] For details see, for instance, J. Lamperti [1966] *Probability: A Survey of the Mathematical Theory,* Benjamin.

For instance, if $S = \{s_1, \ldots, s_N\}$ and the probability of s_i occurring is p_i, then

$$E(\eta) = \sum_{i=1}^{n} \eta(s_i) p_i$$

Suppose there is a function f on the real line such that the probability of η lying between a and b is $\int_a^b f(x)\,dx$, that is,

$$P(\{s \in S \mid \eta(s) \in [a,b]\}) = \int_a^b f(x)\,dx.$$

Then we say that η has the ***probability density function*** f. Clearly, $\int_{-\infty}^{\infty} f(x)\,dx = 1$. Also, one can show that

$$E(\eta) = \int_{-\infty}^{\infty} x f(x)\,dx.$$

The ***variance*** of η is defined by

$$\mathrm{Var}(\eta) = E\left([\eta - E(\eta)]^2\right) = E(\eta^2) - [E(\eta)]^2$$

and the ***standard deviation*** by

$$\sigma(\eta) = \sqrt{\mathrm{Var}(\eta)}.$$

Two random variables, η_1 and η_2, are called ***independent*** if for any two sets A_1, A_2 in the real line $\mathbb{R}$, the events

$$\{s \in S \mid \eta_1(s) \in A_1\} \quad \text{and} \quad \{s \in S \mid \eta_2(s) \in A_2\}$$

are independent. For independent random variables, one has

$$E(\eta_1 \eta_2) = E(\eta_1) E(\eta_2)$$

and

$$\mathrm{Var}(\eta_1 + \eta_2) = \mathrm{Var}(\eta_1) + \mathrm{Var}(\eta_2).$$

(From the definition, $E(\eta_1 + \eta_2) = E(\eta_1) + E(\eta_2)$ is always true.)

The ***law of large numbers*** states that if $\eta_1, \eta_2, \ldots, \eta_n$ are random variables that are independent and have the same mean and variance as η, then

$$E(\eta) = \lim_{n \to \infty} \frac{1}{n} \sum_{i=1}^{n} \eta_i.$$

Part of the theorem is that the right-hand side is a constant. This result sheds light on our intuition that $E(\eta)$ is the average value of η when the experiment is repeated many times. The meaning of the standard deviation is illuminated by ***Tchebysheff's inequality***: If σ is the standard deviation of η,

$$P(\{s \in S \mid |\eta(s) - E(\eta)| \geq k\sigma\}) \leq \frac{1}{k^2}$$

for any number $k > 0$. For example, the probability that η will deviate from its mean by more than two standard deviations is at most $1/4$.

If a random variable η has the probability density function

$$f(x) = \frac{1}{\sqrt{2\pi\sigma^2}} e^{-(x-a)^2/2\sigma^2},$$

we say that η is ***Gaussian***. One can check that $E(\eta) = a$ and $\mathrm{Var}(\eta) = \sigma^2$. If η_1 and η_2 are independent Gaussian random variables, then $\eta_1 + \eta_2$ is Gaussian as well. The ***central limit theorem*** states that if $\eta_1, \eta_2, \ldots$ are independent random variables with mean 0 and variance σ^2, then the probability density function of $(1/\sqrt{n})(\eta_1 + \cdots + \eta_n)$ converges to that of a Gaussian with mean 0 and variance σ^2 as $n \to \infty$. For instance, if measurement errors arise from a large number of independent random variables, the errors can be expected to have a Gaussian distribution. In many problems, however, one does not get a Gaussian distribution for the errors because they arise from nonindependent sources.

Next we show how Gaussian random variables can be used in the study of the heat equation for an infinite rod:

$$v_t = \nu v_{xx}, \qquad -\infty < x < \infty,\ t \geq 0. \tag{2.3.1}$$

Here v represents the temperature in the rod as a function of x and t, and ν represents the conductivity of the rod. If v is given at $t = 0$, then (2.3.1) determines it for $t \geq 0$. If initially $v(x, 0) = \delta(x)$, a delta function at the origin, then the solution of (2.2.1) is given by

$$G(x,t) = \frac{1}{\sqrt{4\pi\nu t}} \exp\left(\frac{-x^2}{4\nu t}\right). \tag{2.3.2}$$

This is the Green's function for the heat equation (see any textbook on partial differential equations).

We can interpret the function (2.3.2) from a probabilistic point of view in two ways as follows.

Method 1 Fix time at t, and place N particles at the origin. Let each of the particles "jump" by sampling the Gaussian distribution with mean zero and variance $2\nu t$. Thus, the probability that a particle will land between x and $x + dx$ is

$$\frac{1}{\sqrt{4\pi\nu t}} \exp\left(\frac{-x^2}{4\nu t}\right) dx.$$

If we repeat this with a large number of particles, we find

$$\lim_{N\to\infty} \frac{\begin{array}{c}\text{number of particles between}\\ x \text{ and } x+dx \text{ at time } t\end{array}}{N\,dx} = \frac{1}{\sqrt{4\pi\nu t}} \exp\left(\frac{-x^2}{4\nu t}\right). \tag{2.3.3}$$

It is convenient to think of each particle as having mass $1/N$, so the total mass is unity.

Method 2 We split up the time interval $[0,t]$ into l pieces, each with length $\Delta t = t/l$, and carry out the procedure in a step-by-step manner. Again, place N particles at the origin. Let them undergo a ***random walk***, that is, the position of the ith particle at time $m\Delta t$ ($i = 1,\ldots,N; m = 1,\ldots,l$) is

$$\begin{aligned} x_i^{m+1} &= x_i^m + \eta_i^m, \\ x_i^0 &= 0, \end{aligned} \tag{2.3.4}$$

where η_i^m are independent Gaussian random variables, each with mean 0 and variance $2\nu\Delta t$. The final displacement of the ith particle is the sum of its displacements, and has a Gaussian distribution with mean 0 and variance $l \times 2\nu\Delta t = 2\nu t$. Thus, the distribution of particles at time t is given again by formula (2.3.3), and methods 1 and 2 are equivalent.

Next consider the solution $v(x,t)$ of the heat equation with given initial data $v(x,0) = f(x)$. The solution is

$$v(x,t) = \int_{-\infty}^{\infty} G(x,x',t) f(x')\,dx', \tag{2.3.5}$$

where

$$G(x,x',t) = \frac{1}{\sqrt{4\pi\nu t}} \exp\left(\frac{-(x-x')^2}{4\nu t}\right).$$

This general solution has a probabilistic interpretation as well. Instead of starting N particles at the origin, start N randomly spaced particles on the line, at positions, say, $x_i^0, i = 1,\ldots,N$, and assign to the ith particle the mass

$$\frac{f(x_i^0)}{N}.$$

Let these particles perform a random walk as in $(2.3.4)_1$, keeping their mass fixed. Then after l steps as in method 2, the expected distribution of mass on the real line approximates (2.3.5).

In this process the total mass of the particles remains constant. This corresponds to the fact that

$$\partial_t \int_{-\infty}^{\infty} v(x,t)\,dx = \nu \int_{-\infty}^{\infty} v_{xx}(x,t)\,dx = 0$$

(assuming $v_x \to 0$ as $x \to \pm\infty$). Of course, one's intuitive feeling that the solutions of the heat equation decay is also correct. Indeed,

$$\partial_t \int_{-\infty}^{\infty} v^2(x,t)\,dx = \int_{-\infty}^{\infty} 2\nu v v_{xx}\,dx = -2\nu \int_{-\infty}^{\infty} (v_x)^2\,dx < 0.$$

The decay of $\int v^2 dx$ (which occurs while $\int v\,dx$ remains constant) is accomplished by spreading. As time advances, the maxima of the solution decay and the variation of the solution decreases. To see intuitively why the integral of v^2 decreases, consider the two functions

$$v_1 = \begin{cases} 2, & -\frac{1}{2} \le x \le \frac{1}{2} \\ 0, & \text{elsewhere} \end{cases} \quad \text{and} \quad v_2 = \begin{cases} 1, & -1 \le x \le 1 \\ 0, & \text{elsewhere} \end{cases}.$$

The function v_2 is more "spread out" than v_1. This is reflected by the calculations

$$\int v_2\,dx = \int v_1\,dx = 2,$$

but

$$\int v_2^2\,dx = 2 \quad \text{and} \quad \int v_1^2\,dx = 4.$$

Note that as time unfolds, the variance of the random walk that is used to construct the solution *increases*, whereas the integral of v^2, which is related to the variance of v, *decreases*. The variance of the random walk increases as the solution spreads out, whereas the integral of v^2 decreases because the range of values assumed by v decreases.

Next consider the heat equation on the halfline $x \le 0$ with boundary condition $v(0,t) = 0$. The Green's function for this problem is

$$G^*(x,x',t) = G(x,x',t) - G(x,-x',t), \tag{2.3.6}$$

where $G(x,x',t)$ is given by $(2.3.5)_2$. That is, G^* satisfies

$$G^*(0,x',t) = 0, \quad G^*(x,x',0) = \delta(x - x'),$$

and

$$\partial_t G^*(x,x',t) = \nu \partial_x^2 G^*(x,x',t).$$

The solution of

$$v_t = \nu v_{xx}, \qquad x \le 0, \;\; t > 0,$$

with

$$v(x,0) = f(x) \quad \text{and} \quad v(0,t) = 0$$

is given in terms of the initial data and the Green's function by

$$v(x,t) = \int_{-\infty}^{\infty} G^*(x,x',t) f(x')\, dx'. \tag{2.3.7}$$

The random walk interpretation for (2.3.6) is obtained by superposing the random walks that generate Green's function for the whole line; namely, the distribution $G^*(x,x',t)$ is obtained by starting N particles with weight $1/N$ at x' and N with weight $-1/N$ at $-x'$, and letting them all walk by, say, method 2. There is an analogous interpretation for (2.3.7), as shown in Figure 2.3.1.

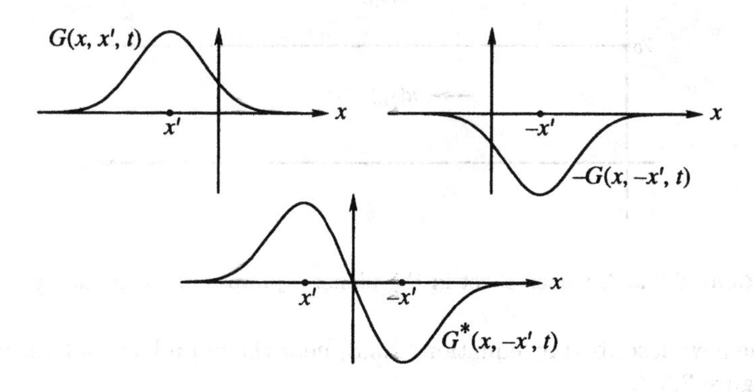

FIGURE 2.3.1. Green's function G^* is obtained by superimposing two random walks.

Random walk methods will now be applied to vortex sheets. Consider flow past an infinite flat plate at rest. In §**2.2** (see Figure 2.2.3), we saw that the flow is

$$u = u(y,t), \qquad v = 0,$$

where

$$u = U \operatorname{erf}(\eta), \quad \text{with} \quad \eta = \frac{y}{\sqrt{4\nu t}}, \tag{2.3.8}$$

and the vorticity is

$$\xi = -\frac{U}{\sqrt{4\nu t}} \exp\left(\frac{-y^2}{4\nu t}\right). \tag{2.3.9}$$

The velocity and vorticity satisfy

$$u_t = \nu u_{yy}, \quad u(0,t) = 0, \quad u(\infty,t) = U, \quad \xi_t = \nu \xi_{yy}. \tag{2.3.10}$$

The boundary conditions for the vorticity are not explicit; they must be determined from the fact that the velocity vanishes on the boundary.

To reconstruct this solution using random walks, we define a ***vortex sheet of intensity*** ξ as follows. As in Figure 2.3.2, it consists of a flow parallel to the x-axis such that u jumps by the amount ξ as y crosses a line, say $y = y_0$; that is,

$$u(y_0+) - u(y_0-) = -\xi.$$

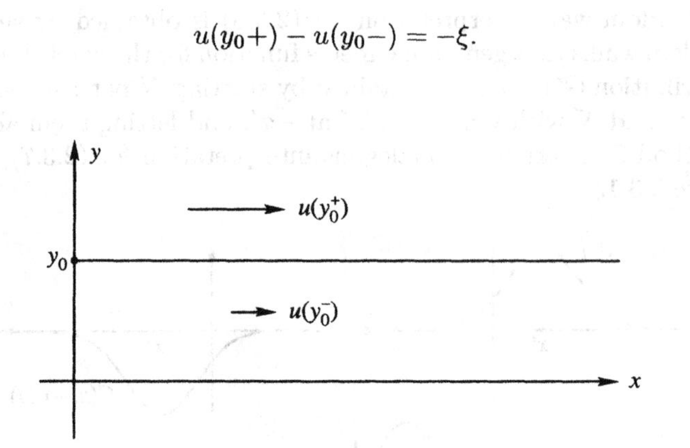

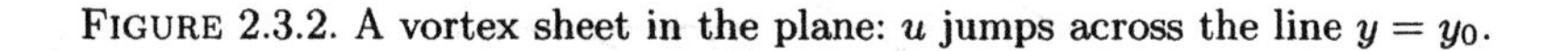

FIGURE 2.3.2. A vortex sheet in the plane: u jumps across the line $y = y_0$.

The flow described by equation (2.3.8) near the boundary looks as shown in Figure 2.3.3.

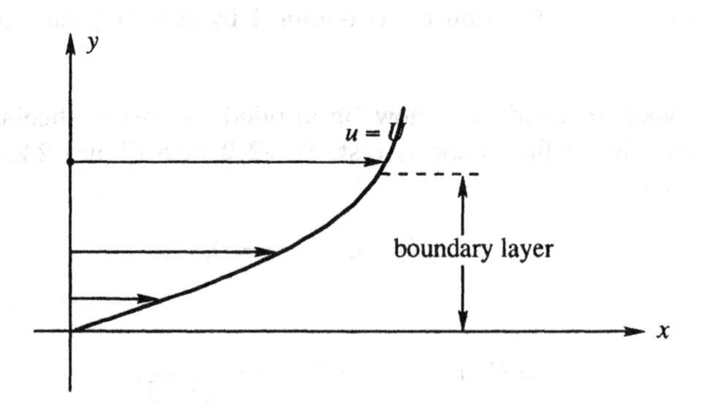

FIGURE 2.3.3. The flow (2.3.8).

As $t \to 0+$, the solution approaches the constant value U for $y > 0$ and yet $u = 0$ on the x-axis; that is, as $t \to 0+$, the solution approaches a vortex sheet on the x-axis with intensity $-U$. We replace this vortex sheet by N "smaller" vortex sheets, each of intensity $-2U/N$. Allow each of these smaller vortex sheets to undergo a random walk in the y-direction of the form

$$y_i^{m+1} = y_i^m + \eta_i^m, \qquad y_i^0 = 0,$$

where η_i^m is chosen from a Gaussian distribution with mean zero and variance $2\nu\Delta t$, where, as in method 2, $\Delta t = t/l, m = 1, 2, \ldots, l$. We claim that for N large the distribution of vorticity constructed this way and the resulting velocity

$$u(y,t) = U + \int_y^\infty \xi(y,t)\, dy \tag{2.3.11}$$

satisfy the correct equations (2.3.10). That ξ and hence u satisfy the heat equation follows from our random walk solution of the heat equation. What requires explanation is why u should vanish on the boundary. To see this, note that on the average, half the vortex sheets are above the x-axis and half below, because they started on the x-axis and the Gaussian distribution has mean zero. Thus,

$$u(0,t) = U + \int_0^\infty \xi(y,t)\, dy,$$

or, in our discrete version, we have on the average

$$u(0,t) = U + \sum_{i=1}^{N/2} \xi_i.$$

But the intensity of the ith vortex sheet is $\xi_i = -2U/N$, and therefore $u(0,t) = 0$. (By symmetry in the x-axis the random walks below the x-axis can be ignored; whenever a vortex sheet crosses the x-axis from above, we can "reflect" it back. On the average, this will balance those vortex sheets that would have crossed the x-axis from below during the course of their random walk. This device can save half the computational effort.)

The random walk method based on vortex sheets will now be generalized to the solution of the Prandtl boundary layer equations:

$$\xi_t + u\xi_x + v\xi_y = \nu\xi_{yy},$$
$$\partial_x u + \partial_y v = 0,$$

with $u = 0 = v$ on the x-axis (the boundary of the region), $y \geq 0$, and $u(\infty) = U$. (We shall see shortly why v cannot be prescribed at ∞.)

The overall flow will be approximated at $t = 0$ by a collection of N vortex sheets of finite width h, extending from $x_i - h/2$ to $x_i + h/2$ with y-coordinate y_i and with strength ξ_i. This approximation is in the same spirit as the point vortex approximation discussed in §**2.1** (see Figure 2.3.4).

To move these sheets, we divide the time $[0, t]$ into l parts of duration $\Delta t = t/l$ and proceed in a step-by-step manner. The advancement in time from time t to $t + \Delta t$ takes place by means of the following algorithm:

i The vortices move according to a discrete approximation to the Euler flow

$$\partial_t \xi + u\xi_x + v\xi_y = 0, \qquad \partial_x u + \partial_y v = 0.$$

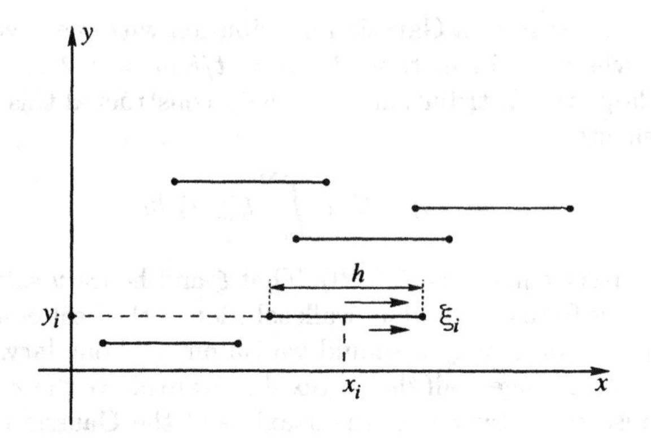

FIGURE 2.3.4. A collection of vortex sheets.

ii Vorticity is added by placing new vortex sheets on the boundary so that the resulting flow has $u = 0 = v$ on the boundary (creation of vorticity).

iii The vortex sheets undergo a random walk as described in the flat plate example (including reflections) to approximate the solution of the equation

$$\xi_t = \nu \xi_{yy}$$

and preserve the boundary conditions $u = 0 = v$.

iv Time is advanced by Δt and one goes back to step i of the procedure, etc., until time t is reached.

Notice that the number of vortex sheets will increase in time; this corresponds to the fact that vorticity is created in boundary layers.

We now give some more details on this procedure. First, we discuss how the vortex sheets move according to the Euler flow. The velocity field satisfies (see §**2.2**):

$$u(x,y) = u(\infty) - \int_y^\infty \frac{\partial u}{\partial y}\, dy = u(\infty) + \int_y^\infty \xi\, dy,$$

or in a discrete version, the velocity of the ith vortex is

$$u(x_i, y_i) = u(\infty) + \sum_j \xi_j, \tag{2.3.12}$$

where the sum is over all vortex sheets such that $y_j > y_i$ and $|x_i - x_j| < h/2$; that is, all vortex sheets whose "shadow" (on the x-axis) engulfs (x_i, y_i); see Figure 2.3.5. Equation (2.3.12) determines the u-component of the velocity field produced by the vortex sheets.

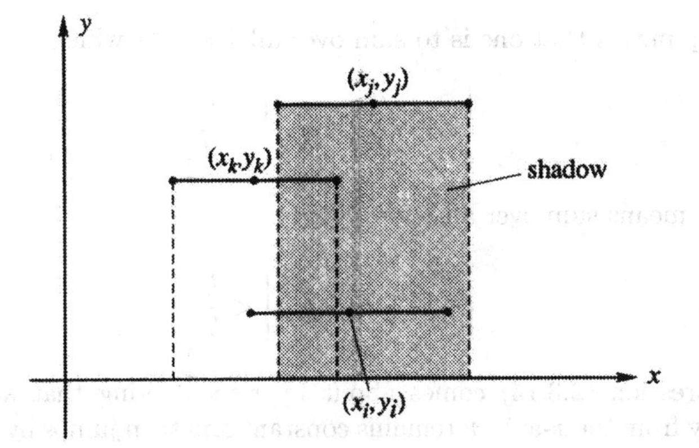

FIGURE 2.3.5. Vortex sheets and their shadows.

From incompressibility and the boundary condition $v(x,0,t)=0$, we find

$$v(x,y,t)=v(x,0,t)+\int_0^y u_x(x,s,t)\,ds=\partial_x\int_0^y u(x,s,t)\,ds.$$

This equation determines v in terms of u (and shows why we could not prescribe v at ∞). A discrete evaluation of y may be given by

$$v(x,y,t)=-\frac{1}{h}\left[\int_0^y u\left(x+\frac{h}{2},s,t\right)ds\right.$$
$$\left.-\int_0^y u\left(x-\frac{h}{2},s,t\right)ds\right]. \quad (2.3.13)$$

A more useful approximation is given by rewriting (2.3.13) directly in terms of the vortex strengths ξ_j, namely,

$$v_i(x_i,y_i,t)=\frac{1}{h}(I_+^i-I_-^i), \quad (2.3.14)$$

where

$$I_+^i(x_i,y_i)=\sum\nolimits_+ y_j^*\xi_j,$$

and

$$I_-^i(x_i,y_i)=\sum\nolimits_- y_j^*\xi_j.$$

with

$$y_j^*=\min(y_j,y_i).$$

Here $\sum_+$ means that one is to sum over all $j \neq i$ for which

$$\left|x_j - \left(x_i + \frac{h}{2}\right)\right| < \frac{h}{2},$$

and $\sum_-$ means sum over all j for which

$$\left|x_j - \left(x_i - \frac{h}{2}\right)\right| < \frac{h}{2}.$$

The expression (2.3.14) comes about by remembering that as we move vertically from the x-axis, u remains constant and then jumps by an amount $-\xi_j$ when a vortex sheet of strength ξ_j is crossed. This leads us directly from (2.3.13) to (2.3.14).

We can summarize all this by saying that in our first step of the algorithm the ith vortex sheet is moved by

$$x_i^{m+1} = x_i^m + u_i \Delta t \qquad y_i^{m+1} = y_i^m + v_i \Delta t, \tag{2.3.15}$$

where u_i is given by (2.3.12) and v_i by (2.3.14). The new velocity field is now determined by the same vortex sheets, but at their new positions. This new velocity field satisfies $v = 0$ on the x-axis by construction, and $u(\infty) = U$. However, because the vortex sheets move, even if $u = 0$ on the x-axis at the beginning of the procedure, it need not remain so. This is another aspect of the main fact we are dealing with: The boundary conditions for the Euler equations, $\mathbf{u} \cdot \mathbf{n} = 0$, are different from those for the Navier–Stokes or Prandtl equations, $\mathbf{u} = \mathbf{0}$.

The purpose of the second step in our procedure is to correct the boundary condition. This may be done as follows: Divide the x-axis into segments of length h, and suppose that at the center P_l of one of these segments $u = u_l \neq 0$. Then at P_l place one (or more) vortex sheets, the sum of whose intensities is $2u_l$. This will guarantee that (on the average) $u = 0$ on the x-axis in the new flow.

In the third step in our procedure, we add a random y component to positions (x_i, y_i) of the existing (as well as of the newly created) vortex sheets:

$$x_i^{m+1} = x_i^m + u_i \Delta t \qquad y_i^{m+1} = y_i^m + v_i \Delta t + \eta_i^m, \tag{2.3.16}$$

where η_i^m are Gaussian random variables with mean 0 and variance $2\nu\Delta t$.

Intuitively, the vortex sheets move about in ideal flow together with a random y component simulating viscous diffusion. Vortex sheets newly created to accommodate the boundary conditions diffuse out from the boundary by means of the random component and then get swept downstream by the main flow. This mechanism then has all the features a boundary layer should have as we discussed at the beginning of §**2.2**.[9]

Consider the problem of flow past a semi-infinite flat plate, situated on the positive x-axis (Figure 2.3.6). Far enough along the positive x-axis, say at $x = x_2$, most of the vortex sheets that move downstream are replaced by vortex sheets coming from the left. Thus, there is little need to create much vorticity at x_2.

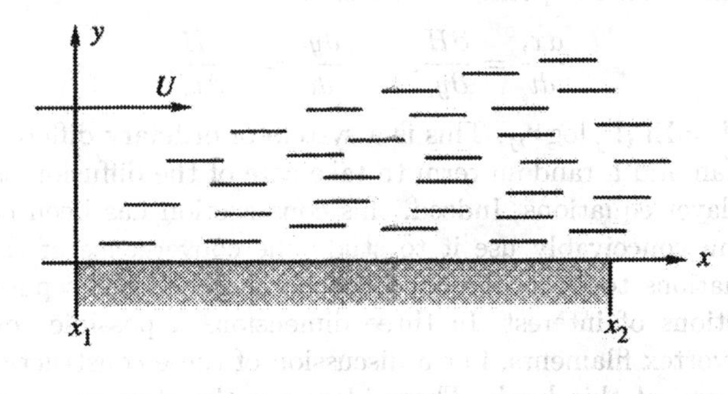

FIGURE 2.3.6. Vortex sheets get created and swept downstream.

However, at x_1 (called the ***leading edge***) any vortex sheet is immediately swept downstream by the flow with no replacement. Thus, we are constantly forced to create more vortex sheets at $x = x_1$ to satisfy the no-slip condition. One can see that most of the vorticity in the flow is created at the leading edge.

In a time Δt, how far does a vortex sheet move? In the x direction the displacement is proportional to (Δt); in the y direction the displacement is $(\Delta t)v_i + \eta_i^m$. The standard deviation tells us the length of the average "jump." Thus,

$$\text{average jump in } y \text{ direction} \cong \sqrt{\text{var}} \sim \sqrt{\Delta t}.$$

The term $(\Delta t)v_i$ is also proportional to $\sqrt{\Delta t}$, as can be seen from equation (2.3.14) and the fact that I_+ and I_- contain ys as factors. Suppose the flow is stationary; we may regard t as a parameter and we may eliminate

[9]The generation of vorticity is discussed is G. K. Batchelor [1967] *An Introduction to Fluid Mechanics*, Cambridge Univ. Press; J. Lighthill [1963] "Introduction to Boundary Layer Theory", in *Laminar Boundary Layers*, edited by L. Rosenhead, Oxford Univ. Press. The vortex sheet model is due to A. J. Chorin, *J. Comp. Phys.* **27** [1978], 428. Some theoretical aspects are found in A. J. Chorin, T. J. R. Hughes, M. J. McCracken, and J. E. Marsden, *Comm. Pure. Appl. Math.* **31** [1978], 205; C. Anderson, *J. Comp. Phys.*, **80** [1989], 72; C. Marchioro and M. Pulvirenti, *Vortex Methods in 2-Dimensional Fluid Mechanics*, Springer-Verlag, [1984]; K. Gustafson and J. Sethian, *Vortex Flows*, SIAM Publications [1991].

it to find

$$\text{vertical displacement} \sim \sqrt{\text{horizontal displacement}}.$$

This shows that the structure of the boundary on a semi-infinite flat plate is parabolic, as was derived earlier.

Finally, we ask, are there more general constructions available, similar to this vortex sheet construction? For Euler's equations, we have already described the model of point vortices satisfying

$$\frac{dx_i}{dt} = \frac{\partial H}{\partial y_i}, \qquad \frac{dy_i}{dt} = -\frac{\partial H}{\partial x_i},$$

where $4\pi H = \Sigma\Gamma_i\Gamma_j \log r_{ij}$. This is a system of ordinary differential equations; we can add a random term to take care of the diffusion as with the boundary layer equations. Indeed, this construction has been carried out and one can conceivably use it to study the convergence of the Navier–Stokes equations to Prandtl's equations, boundary layer separation, and other questions of interest. In three dimensions a possible construction might use vortex filaments, but a discussion of these constructions is outside the scope of this book. These ideas are the starting points for the construction of practical numerical algorithms.

There is a way of writing the above general scheme for solving the Navier–Stokes equations that sheds light on the mathematical structure of this and related methods.[10] The methods are related to some basic facts about algorithms and our purpose is just to provide a hint about them.

Consider first a differential equation

$$\dot{x} = X(x) \tag{2.3.17}$$

on $\mathbb{R}^n$ (eventually replaced by a more general space). To solve this equation, a computer would move a point forward during a time step $\Delta t = \tau$ according to some rule. More precisely, an ***algorithm*** is a collection of maps $F_\tau : \mathbb{R}^n \to \mathbb{R}^n$. The associated iterative process is denoted $x_{k+1} = F_\tau(x_k)$. To be ***consistent*** with (2.3.17) one requires

$$\left.\frac{d}{d\tau}F_\tau(x)\right|_{\tau=0} = X(x). \tag{2.3.18}$$

Under some additional hypotheses one can establish ***convergence*** of the algorithm, namely,

$$\lim_{n\to\infty}(F_{t/n})^n x_0 = x(t), \tag{2.3.19}$$

[10]Some additional references, in addition to those in the preceding footnote, are A. J. Chorin, *J. Fluid Mech.* **57** [1973], 785–796, J. E. Marsden, *Bull. Amer. Math. Soc.* **80** [1974], 154–158, G. Benfatto and M. Pulvirenti, Convergence of the Chorin-Marsden product formula in the half-plane, *Comm. Math. Phys.* **106** [1986], 427–458, and J. E. Marsden, *Lectures on Mechanics*, Cambridge University Press, [1992].

where $x(t)$ is the solution of (2.3.17) with initial condition x_0. Of course, questions of rate of convergence and efficiency of the algorithm are important in practice, but are not discussed here.

An important simple case is the following. Consider solving the linear system

$$\dot{x} = Ax + Bx, \tag{2.3.20}$$

where A and B are $n \times n$ matrices. The solution is

$$x(t) = e^{t(A+B)}x_0, \tag{2.3.21}$$

where

$$e^C = I + C + \tfrac{1}{2}C^2 + \tfrac{1}{3!}C^3 + \cdots$$

However, it may be easier to solve the equation by breaking it into two pieces: $\dot{x} = Ax$ and $\dot{x} = Bx$ and solving these successively. Doing so leads to the algorithm

$$F_\tau x = e^{\tau A}e^{\tau B}x,$$

and yields the basic formula

$$e^{t(A+B)} = \lim_{n \to \infty} (e^{tA/n}e^{tB/n})^n \tag{2.3.22}$$

(sometimes called the ***Lie–Trotter product formula***). More generally, consider solving a *nonlinear* equation

$$\dot{x} = X(x) + Y(x). \tag{2.3.23}$$

Let $H_t(x_0)$ denote the solution of (2.3.23) with initial condition x_0, so we thereby define a map $H_t : \mathbb{R}^n \to \mathbb{R}^n$, called the ***flow map***. Let K_t and L_t be the corresponding flow maps for $\dot{x} = X(x)$ and $\dot{x} = Y(x)$. Then (2.3.22) generalizes to

$$H_t(x_0) = \lim_{n \to \infty} (K_{t/n} \circ L_{t/n})^n(x_0), \tag{2.3.24}$$

where the power n indicates repeated composition.

It is tempting to apply (2.3.24) to the Navier–Stokes equations, where K_t is the flow of the Stokes equation and L_t the flow of the Euler equations. First of all, consider the case of regions with *no boundary* (e.g., this occurs for the case of spatially periodic flows in the plane or space). In this case, the method works quite well and was used by Ebin and Marsden in 1970 to show the convergence in a suitable sense of the solutions of the Navier–Stokes equations to the Euler equations as the viscosity $\nu \to 0$.

If a boundary is present, we need to modify the scheme because of the boundary conditions, as has been explained. Translated into a product formula, the modified scheme reads as follows.

$$H_t(\mathbf{u}) = \lim_{n\to\infty} (K_{t/n} \circ \Phi_{t/n} \circ L_{t/n})^n \mathbf{u}. \tag{2.3.25}$$

Here,

- $\mathbf{u}$ is a divergence-free vector field, $\mathbf{u} = \mathbf{0}$, on the boundary $\partial\Omega$ of the region Ω in question;
- L_t is the flow of the Euler equation (boundary conditions $\mathbf{u}$ parallel to $\partial\Omega$), which may be solved approximately by a vortex method;
- K_t is the flow of the Stokes equation (boundary conditions $\mathbf{u} = \mathbf{0}$ on $\partial\Omega$), solved for example by a random walk procedure on vorticity;
- Φ_t is a "vorticity creation operator" that maps a $\mathbf{u} \parallel \partial\Omega$ to a $\mathbf{u} = \mathbf{0}$ on $\partial\Omega$ by adding on a suitably constructed vorticity field to $\mathbf{u}$ whose backflow cancels $\mathbf{u}$ on $\partial\Omega$; and
- H_t is the flow of the Navier–Stokes equations.

Convergence of (2.3.25) in reasonable generality is not yet proven, although special cases are given in the cited references. What formula (2.3.25) does is to make explicit the intuitive idea that vorticity is created on the boundary because of the difference in the boundary conditions between the Euler and Navier–Stokes equations, and that if the Reynolds number is high this vorticity is swept downstream by the Euler flow to form a structured or turbulent wake. The references should be consulted for details.

2.4 Remarks on Stability and Bifurcation

If R is the Reynold's number, then limit $R \to 0$ corresponds to slow or viscous flow. The other extreme, the limit $R \to \infty$ concerns very fast or only slightly viscous flow. The situation for intermediate values of R involves many interesting transitions, or changes of flow pattern, and is the subject of much research by mathematicians and engineers alike. Here we shall give a few informal comments using the subject of nonlinear dynamics, complementing topics already discussed in this chapter. The concept of *stability* plays a central role in this discussion, so we begin with this concept.

Let P be $\mathbb{R}^n$ (more generally a Banach or Hilbert space of functions), and let F_t denote a flow of a differential equation $\dot{x} = X(x)$ on P; that is,

$$\frac{d}{dt}F_t(x) = X(F_t(x), t),$$

where $x \in P$ and $t \geq 0$. Assume that we have a good existence and uniqueness theorem. If X is independent of t, we say it is ***autonomous***, and if x_0 is such that $X(x_0) = 0$, we call x_0 a ***fixed point*** of X. It follows from the initial condition $F_0(x_0) = x_0$ and uniqueness of solutions that x_0 is a *fixed point* of X, *i.e.*,

$$F_t(x_0) = x_0.$$

Definition *A point x_0 is an **asymptotically stable fixed point** of the vector field X if there is a neighborhood $U \subset P$ of x_0 such that if $x \in U$, then $F_t(x) \to x_0$ as $t \to \infty$. In the sequel we shall refer to such a point simply as a **stable point**. If one can choose $U = P$, we say that x_0 is **globally stable**. See Figure 2.4.1.*

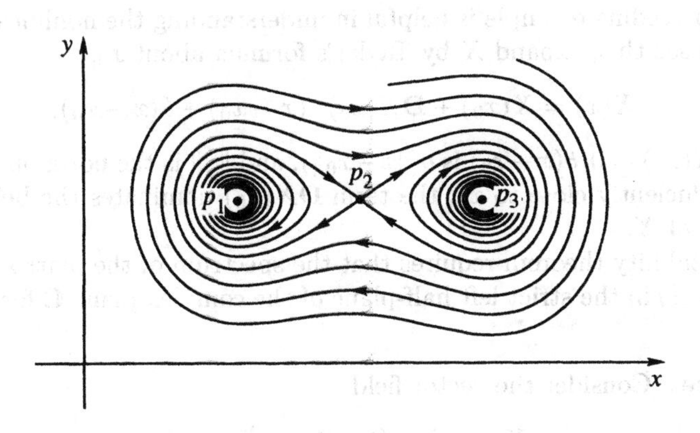

FIGURE 2.4.1. A system with two asymptotically stable points.

Next we give a basic stability result of Liapunov[11] that is based on computing the spectrum of the linearization.

Liapunov Stability Theorem *Let x_0 be a fixed point of a smooth autonomous vector field X on $P = \mathbb{R}^n$. Then x_0 is a stable point if the eigenvalues of the $n \times n$ matrix $\mathbf{D}X(x_0)$ of partial derivatives of X have real parts < 0.*

Example Suppose we have a linear autonomous system, that is, $X = A$, an $n \times n$ real matrix, which we assume for simplicity is diagonalizable. Then the solution of

$$\frac{d}{dt}F_t(x) = A \cdot F_t(x), \quad F_0(x) = x$$

---

[11]For the proof, see for example, M. Hirsch and S. Smale [1974] *Differential Equations, Dynamical Systems, and Linear Algebra,* Academic Press, or the book by Abraham, Marsden, and Ratiu listed in the Preface.

is $F_t(x) = e^{tA}x$, where $e^{tA} = \sum_{i=o}^{\infty} t^i A^i / i!$. This series is absolutely convergent. Transforming to coordinates that diagonalize A, we get

$$A = \operatorname{diag}(\lambda_1, \ldots, \lambda_n)$$

and so

$$e^{tA} = \operatorname{diag}(e^{t\lambda_1}, \ldots, e^{t\lambda_n})$$

where $\lambda_1, \ldots, \lambda_n$ are the eigenvalues of A (they occur in conjugate pairs because A is real). This *indicates* why the stability theorem is true, namely, if $\operatorname{Re}(\lambda_i) < 0, i = l, \ldots, n$, then the matrix e^{tA} converges to the zero matrix as $t \to \infty$. Observe that for the *linear* case, consideration of neighborhoods is unnecessary. ♦

The preceding example is helpful in understanding the nonlinear case as well. To see this, expand X by Taylor's formula about x_0:

$$X(x) = X(x_0) + \mathbf{D}X(x_0) \cdot (x - x_0) + \theta(x - x_0),$$

where $X(x_0) = 0, \theta(x - x_0)$ is $o(\|x - x_0\|)$, and $\|\cdot\|$ is the norm on P. Thus, for x sufficiently close to x_0, the term $\mathbf{D}X(x_0)$ dominates the behavior of the flow of X.

The stability theorem requires that the spectrum of the matrix $\mathbf{D}X(x_0)$ lies *entirely* in the strict left half-plane of the complex plane $\mathbb{C}$ for x_0 to be stable.

Exercise Consider the vector field

$$X_\mu(x, y) = (y, \mu(1 - x^2)y - x)$$

on $\mathbb{R}^2$. Determine if the fixed point $(0, 0)$ is stable or unstable for various values of μ. ♦

We shall now examine how these concepts enable us to study the stability of solutions of the Navier–Stokes equations. Consider the example of flow in a pipe and assume v_0 is a stationary solution of the Navier–Stokes equations, for example, the Poiseuille solution (Exercise 1.3-3). We are interested in what happens when v_0 is perturbed, that is, when $v_0 \to v_0 + \delta v$. Note that the disturbed solution of the Navier–Stokes equations, $v_0 + \delta v$, will not be a stationary solution. This is part of our notion of stability, that is, v_0 is stationary corresponding to v_0 being a fixed point of a vector field on the space of all possible solutions of the Navier–Stokes equations, which we shall denote by P. For the present example, P would consist of the set of all divergence-free velocity fields satisfying the appropriate boundary conditions. The Navier–Stokes equations determine the dynamics on this space P of velocity fields. We write

$$\frac{dv}{dt} = X(v), \qquad v \text{ a curve in } P,$$

where X does not depend on time unless the boundary conditions or body force are time dependent. In this abstract notation, the Navier–Stokes equations, boundary conditions, divergence-free condition, and so on, look like an ordinary differential equation. In the present circumstances X is an unbounded operator. However, we can still apply the stability concepts for dynamical systems on $\mathbb{R}^n$, due to the following conditions:

(1) The preceding equation really does determine the dynamics on P. This follows from the short t-interval existence and the uniqueness theorem for the Navier–Stokes equations.

(2) The stability theorem really works for the Navier–Stokes equations.[12]

For condition (2), $\mathbf{D}X$ is obtained via differentiation and $\mathbf{D}X(v_0)$ is a linear differential operator. Its spectrum will consist of infinitely many eigenvalues or it may be continuous. For v_0 to be stable the entire spectrum must lie in the left half-plane of $\mathbb{C}$. For example, flow in a pipe and Couette flow are stable (in fact globally stable) if the Reynolds number is not too big.

Sometimes one is interested in the loss of stability of flows as the Reynolds number R is increased. In general, any such qualitative change in the nature of a flow is called a ***bifurcation.*** In this regard we consider X to be parametrized by R and study the behavior of the spectrum of $\mathbf{D}X(v_0)$ as a function of this parameter. As R is increased, we anticipate that conjugate pairs of eigenvalues of $\mathbf{D}X(v_0)$ may drift across the imaginary axis. In this case stability is lost and an oscillation develops.

A major result dealing with this situation is given by the following theorem of Poincaré, Andronov, and Hopf.[13]

Hopf Bifurcation Theorem *(Nontechnical version) Let X_μ be a vector field (depending on a parameter μ) possessing a fixed point x_0 for all μ. Assume the eigenvalues of $\mathbf{D}X_\mu(x_0)$ are in the left half-plane of $\mathbb{C}$ for $\mu < \mu_0$, where μ_0 is fixed. Assume that as μ is increased, a single conjugate pair $\lambda(\mu), \overline{\lambda(\mu)}$ of eigenvalues crosses the imaginary axis with nonzero speed at $\mu = \mu_0$ (Figure 2.4.2). Then there is a family of closed orbits of X near μ_0. If the point x_0 is stable for X_{μ_0}, then the closed orbits appear for $\mu > \mu_0$ and are stable. For each $\mu > \mu_0$, near μ_0, there is one corresponding stable closed orbit and its period is approximately equal to* $\mathrm{Im}(\lambda(\mu_0)/2\pi)$.

---

[12] See for example, D. Joseph, *Stability of Fluid Motion,* I, II, Springer-Verlag and for the Euler equations, V. Arnold and B. Khersin, *Topological Methods in Hydrodynamics*, Springer-Verlag, 1997, and references therein.

[13] References are J. E. Marsden and M. McCracken *The Hopf Bifurcation,* Springer Applied Mathematics Series, Vol. 19 [1976]; D. Sattinger *Lectures on Stability and Bifurcation Theory,* Springer Lecture Notes **309** [1973]. Books that approach this subject from a more physical point of view are C. C. Lin, *The Theory of Hydrodynamic Stability*, Cambridge [1955], and S. Chandrasekhar *Hydrodynamic and Hydromagnetic Stability*, Oxford [1961].

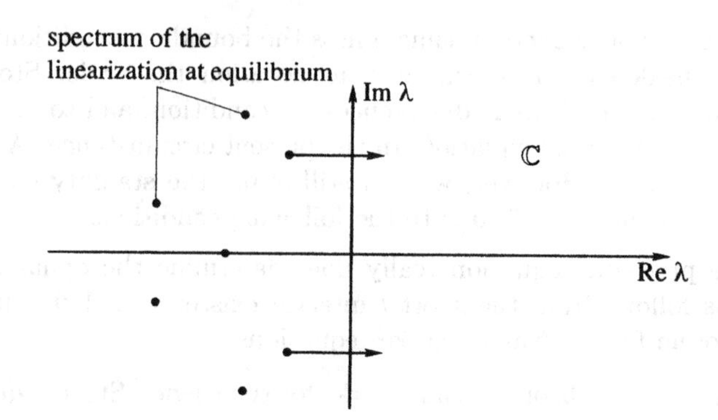

FIGURE 2.4.2. A pair of eigenvalues crosses the imaginary axis in the Hopf bifurcation.

Roughly speaking, what this means is that when stability is lost, a stable point is replaced by a stable closed orbit. Translated to the case of fluid mechanics this means that a stationary solution (fixed point) of the Navier–Stokes equations is replaced by a periodic solution (stable closed orbit). For example, consider the case of flow around a cylinder. As R is increased, the stationary solution becomes unstable and goes to a stable periodic solution—the wiggly wake depicted in Figure 2.4.3. Such a development of periodic oscillations is the main content of the Hopf bifurcation theorem. It applies to a wide variety of physical and biological phenomena, and this ubiquity leads one to strongly suspect it is the mechanism underlying even complex fluid dynamical phenomena like the singing of transmission lines in a strengthening wind.

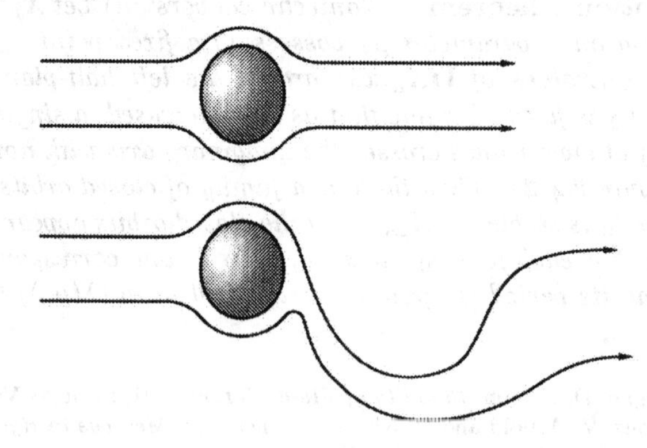

FIGURE 2.4.3. The development of patterns like this is related to bifurcation, stability loss and symmetry breaking.

We say that a ***bifurcation*** has occurred when R reaches a critical value where stability is lost and is replaced by oscillations. The term is used generally when sudden qualitative changes occur. In Hopf's theorem the stability analysis fails at μ_0 and stability must be determined by "higher-order" analysis; that is, the stability theorem, based on first-order analysis just fails at $\mu = \mu_0$. How to do this is discussed in the cited references. One should also note that in highly symmetrical situations such a Couette flow, the hypotheses of Hopf's theorem (simplicity of the eigenvalues, or the nonzero speed assumption) can fail. In this context the theory of bifurcation with symmetry is needed.[14]

As R is increased, additional bifurcations and more complex motions can occur. It is still not clear how one might set useful information about turbulent flows from this approach, but for some complex flows like Taylor-Couette flow the approach has been very successful.

---

[14]See Marsden and Hughes [1994]; M. Golubitsky, I. Stewart and D. Schaeffer *Symmetry and Groups in Bifurcation Theory,* Vol. II, Springer-Verlag [1988]; and P. Chossat and G. Iooss, *The Taylor-Couette Problem*, Springer Applied Math. Sciences Series [1991] for further information and references.

3

Gas Flow in One Dimension

In this chapter we discuss compressible flow in one dimension. In the first section we develop the geometry of characteristics and in the second we introduce the notion of a weak solution and the entropy condition for shocks. In the third section we discuss the Riemann problem, that is, a flow problem with particular discontinuous initial data. A general construction, due to Glimm, which uses the solution of Riemann problems to produce solutions of arbitrary problems, is then presented. This construction is the basis of both some existence proofs and some methods of numerical computation in gas dynamics. In the final section we generalize the discussion to the flow of a gas that allows chemical energy release, such as occurs in combustion.

3.1 Characteristics

One-dimensional isentropic flow with an equation of state $p = p(\rho)$ is described by the following equations derived in §**1.1**:

$$\left.\begin{aligned} u_t + uu_x &= -\frac{p_x}{\rho}, \\ \rho_t + u\rho_x + \rho u_x &= 0. \end{aligned}\right\} \tag{3.1.1}$$

Define $c = \sqrt{p'(\rho)}$ (assuming $p'(\rho) > 0$) and note that $p_x = p'(\rho)\rho_x$ by the chain rule. Thus, the first equation in (3.1.1) becomes

$$u_t + uu_x = -c^2\frac{\rho_x}{\rho}.$$

Before proceeding, let us explain why c is called the ***sound speed***. Sound can be viewed as a small disturbance in an otherwise motionless gas, due to density changes. Let

$$u = u' \quad \text{and} \quad \rho = \rho_0 + \rho',$$

where u' and ρ' are small and ρ_0 is constant. Correspondingly, neglecting products of small quantities, we have

$$p = p_0 + \frac{\partial p}{\partial \rho}\rho' = p_0 + c^2\rho' \quad \text{and} \quad c^2 = \frac{\partial p}{\partial \rho}(\rho_0) = \text{constant}.$$

Substitution of these expressions in (3.1.1) gives

$$u'_t + u'u'_x = -c^2\frac{\rho'_x}{\rho} \quad \text{and} \quad \rho'_t + u\rho'_x + \rho u'_x = 0.$$

To first order (i.e., neglecting again squares of small quantities) these become the linearized equations

$$\rho_0 u'_t = -c^2\rho'_x \quad \text{and} \quad \rho'_t + \rho_0 u'_x = 0.$$

Eliminating u' by differentiating the first equation in x and the second in t, we get

$$\rho'_{tt} = c^2\rho'_{xx},$$

which is the wave equation. A basic fact about this equation is that the general solution is

$$\varphi(x + ct) + \psi(x - ct),$$

that is, small disturbances ("sound waves") propagate with speed c.

If we define

$$A = \begin{bmatrix} u & \rho \\ c^2/\rho & u \end{bmatrix},$$

then (3.1.1) becomes

$$\begin{pmatrix} \rho \\ u \end{pmatrix}_t + A\begin{pmatrix} \rho \\ u \end{pmatrix}_x = 0. \tag{3.1.2}$$

Notice that the eigenvalues of A are $u + c$ and $u - c$. Equation (3.1.2) is a special case of a ***first-order quasilinear hyperbolic system in one dimension***, that is, a special case of the system

$$\mathbf{u}_t + A(x,t,\mathbf{u})\mathbf{u}_x = \mathbf{B}(x,t,\mathbf{u}), \tag{3.1.3}$$

where $\mathbf{u} = \mathbf{u}(x, t)$ is an n-component vector function of x and t, and A is an $n \times n$ matrix function of x, t, and $\mathbf{u}$ that has n distinct real eigenvalues and hence n linearly independent eigenvectors (i.e., as a real matrix, A is diagonalizable). The nonlinearity of (3.1.3) is reflected in the dependence of A or $\mathbf{B}$ on the unknown vector $\mathbf{u}$. We shall now examine systems of the form (3.1.3).

Example 1 Consider the single linear equation

$$v_t - v_x = 0. \tag{3.1.4}$$

Given initial data $v(x, 0) = f(x)$, the solution is

$$v(x, t) = f(x + t).$$

Thus, v is a wave, with shape f, traveling with unit speed to the left on the x-axis. On the lines $x + t =$ constant, v is constant. Thus, we can think of information from the initial data as being propagated along these lines. These lines are called the ***characteristics*** of equation (3.1.4). If initial data are given on a curve C that is transverse to the characteristics (i.e., nowhere tangent to them), then (3.1.4) is solved by setting $v(x_0, t_0) =$ the value of the initial data on the curve C at the point where C intersects the characteristic through (x_0, t_0). See Figure 3.1.1. If initial data are given on

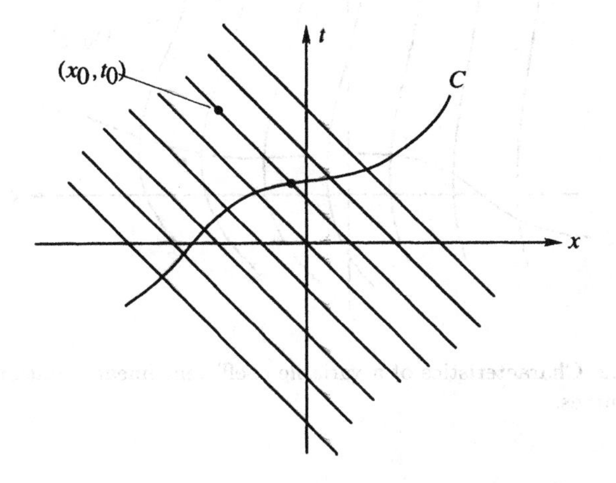

FIGURE 3.1.1. Characteristics intersecting the curve C.

a curve not transverse to the characteristics, such as on a characteristic itself, then the equation does not necessarily have a solution. ◆

Example 2 Consider the general single linear homogeneous equation

$$v_t + a(x, t)v_x = 0. \tag{3.1.5}$$

This time the ***characteristics*** are the curves $t = t(s), x = x(s)$, satisfying

$$\frac{dt}{ds} = 1, \qquad \frac{dx}{ds} = a(x,t). \tag{3.1.6}$$

If $a(x,t)$ is smooth, these curves exist (at least locally) and never intersect without being coincident, by the existence and uniqueness theorem for ordinary differential equations. Let us verify that, as in the first example, solutions are constant along characteristics. Indeed, if $x = x(s), t = t(s)$ is a characteristic and v satisfies (3.1.5), then by the chain rule,

$$\frac{d}{ds}v(x(s),t(s)) = v_x\frac{dx}{ds} + v_t\frac{dt}{ds} = v_t + a(x,t)v_x = 0.$$

Thus, if initial data is given on a curve C that is transverse to the characteristics, we can assert as before that $v(x_0,t_0)$ = the initial data at point P in Figure 3.1.2. If the data are given on a curve that is not transverse to the characteristics, a solution does not necessarily exist. ♦

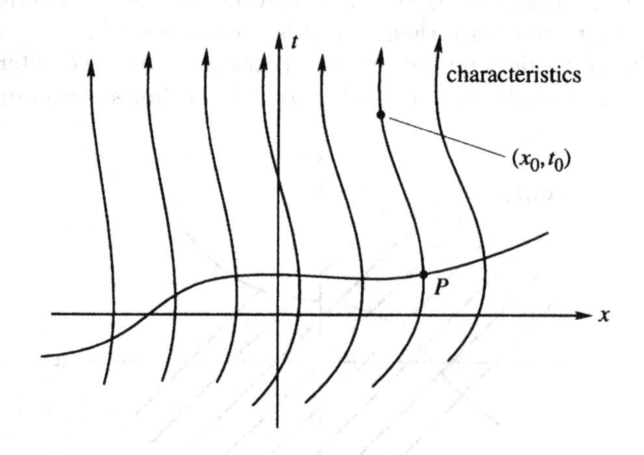

FIGURE 3.1.2. Characteristics of a variable coefficient linear equation are nonintersecting curves.

Example 3 Consider the generalization of (3.1.5) to the inhomogeneous equation

$$v_t + a(x,t)v_x = b(x,t), \tag{3.1.7}$$

and define the characteristics by the same equations (3.1.6). This time

$$\frac{d}{ds}v(x(s),t(s)) = b(x,t),$$

and so, as the right-hand side is known,

$$v(x(s), t(s)) = v(x(s_1), t(s_1)) + \int_{s_1}^{s} b(x(\alpha), t(\alpha))\, d\alpha,$$

which again determines v off any curve C if initial data is given on C and C is transverse to the characteristics. ♦

Example 4 As a *nonlinear* example, consider the equation

$$u_t + uu_x = 0. \tag{3.1.8}$$

Let us search for curves $x(s), t(s)$ along which u is constant. By the chain rule,

$$\frac{d}{ds}u(x(s), t(s)) = u_x \frac{dx}{ds} + u_t \frac{dt}{ds}.$$

Thus, we should choose

$$\frac{dt}{ds} = 1, \qquad \frac{dx}{ds} = u.$$

Thus, the characteristics depend on u. Along the curve defined by $dt/ds = 1$ and $dx/ds = u$, we have $du/ds = 0$; that is, u is constant. The situation is similar to the one in the linear case, except for the following crucial fact: characteristics can now intersect. This is easy to see directly; if $u =$ constant on a characteristic, then $dt/ds = 1, dx/ds = u$ is a straight line, and such lines issuing from different points can indeed intersect (see Figure 3.1.3). It is also easy to see in principle why the nonlinear case differs from the linear case. In the linear case, the characteristics are determined by two ordinary differential equations

$$\frac{dt}{ds} = 1, \qquad \frac{dx}{ds} = a(x, t).$$

If $a(x, t)$ is reasonably smooth, the solution of these equations through a given point (x_0, t_0) is unique, and thus the characteristics cannot intersect. In the nonlinear case, the characteristics are determined by three equations (in the special case above, $dt/ds = 1, dx/ds = u, du/ds = 0$). The solution of these equations through (x_0, t_0, u_0) is still unique, but the characteristics in the (x, t) plane are the projections of the three-dimensional solutions of the system on that plane, and thus they may indeed intersect. When such intersection occurs, our method of solution by following characteristics breaks down and the solution is no longer uniquely determined. We shall resolve this problem in the next section. ♦

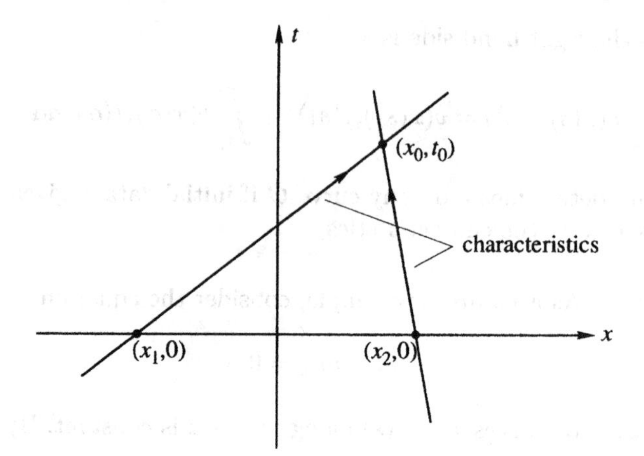

FIGURE 3.1.3. The characteristics of a nonlinear equation may intersect.

Now we return to the system (3.1.3) and define its characteristics. Setting $\mathbf{B} = \mathbf{0}$ for simplicity, the rate of change of $\mathbf{u}$ along a curve $(x(s), t(s))$ is

$$\frac{d\mathbf{u}}{ds} = \mathbf{u}_t \frac{dt}{ds} + \mathbf{u}_x \frac{dx}{ds} = \left[-A(x,t,\mathbf{u}) \frac{dt}{ds} + \frac{dx}{ds} \right] \mathbf{u}_x. \tag{3.1.9}$$

We define a ***characteristic*** to be a curve with the following property: if data are given on that curve, the differential equation does *not* determine the solution at any point *not* on the characteristic. (The data may also fail to be consistent, but we do not make this fact a part of the definition.) One can readily check that all the examples of characteristics we have given meet this definition. If the characteristic is not parallel to the x-axis, this means we cannot determine $\mathbf{u}_x$ from knowledge of $\mathbf{u}$ on the curve. From (3.1.9) this will happen if

$$-A(x,t,u)\frac{dt}{ds} + \frac{dx}{ds} I$$

is a singular matrix, that is, if

$$\frac{dt}{ds} = 1 \quad \text{and} \quad \frac{dx}{ds} = \lambda(x,t,\mathbf{u}), \tag{3.1.10}$$

where $\lambda(x, t, \mathbf{u})$ is an eigenvalue of $A(x, t, \mathbf{u})$. Notice that there are n characteristics through each point and that the characteristics depend on $\mathbf{u}$.

If a change of variables is made on the dependent variable $\mathbf{u}$ that has a nonzero Jacobian, then A is changed by a similarity transformation, and thus the eigenvalues and hence the characteristics are unchanged. If a change of variables is made on the independent variables (x, t) (with nonzero Jacobian), then this change maps characteristics onto characteristics of the transformed problem.

If C is a curve transverse to all the characteristics, then

$$-A\frac{dt}{ds}+\frac{dx}{ds}$$

will be invertible, and from (3.1.9),

$$\mathbf{u}_x=\left[-A\frac{dt}{ds}+\frac{dx}{ds}\right]^{-1}\mathbf{u}_s,$$

so we can propagate $\mathbf{u}$ off C. Thus, it is reasonable to expect (and indeed it is true) that initial data on C uniquely determine a solution near C. (We have to say "near" because, as in Example 4, characteristics for the same eigenvalue can intersect.) In this general case we cannot expect $\mathbf{u}$ to be constant along characteristics. However, we might seek functions $f_1,\ldots,f_n$ of $\mathbf{u}$ that are constant along the characteristics associated with the eigenvalues $\lambda_1,\ldots,\lambda_n$. Indeed, suppose we can find a function f associated with an eigenvalue λ with the property that

$$A^T\frac{\partial f}{\partial \mathbf{u}}=\lambda\frac{\partial f}{\partial \mathbf{u}}, \tag{3.1.11}$$

that is, $\partial f/\partial \mathbf{u}$ is an eigenvector of the transpose of A. We claim that f is constant along the corresponding characteristic. Such functions f are called the ***Riemann invariants*** of the equation. To prove our contention, write out (3.1.11) in components:

$$\sum_{k=1}^{n}A_{ki}\frac{\partial f}{\partial u_k}=\lambda\frac{\partial f}{\partial u_i}, \tag{3.1.12}$$

where $\mathbf{u}=(u_1,\ldots,u_n)$. Now differentiate f along a curve satisfying (3.1.10):

$$\begin{aligned}\frac{\partial f}{\partial s}&=\sum_{k=1}^{n}\frac{\partial f}{\partial u_k}\frac{\partial u_k}{\partial s}=\sum_{k=1}^{n}\frac{\partial f}{\partial u_k}\left[\frac{\partial u_k}{\partial t}+\frac{\partial u_k}{\partial x}\frac{dx}{ds}\right]\\&=\sum_{k=1}^{n}\left[\frac{\partial f}{\partial u_k}\left[-\sum_{i=1}^{n}A_{ki}\frac{\partial u_i}{\partial x}+\lambda\frac{\partial u_k}{\partial x}\right]\right]\\&=\sum_{k=1}^{n}-\sum_{i=1}^{n}A_{ki}\frac{\partial f}{\partial u_k}\frac{\partial u_i}{\partial x}+\sum_{k=1}^{n}\lambda\frac{\partial f}{\partial u_k}\frac{\partial u_k}{\partial x}.\end{aligned}$$

This is zero by virtue of (3.1.12).

Thus, if n functions $\mathbf{f}=(f_1,\ldots,f_n)$ are found that are constant along characteristics, we may invert them to express $\mathbf{u}$ in terms of the $f's$ and hope in this way to be able to determine explicitly the characteristics in terms of the initial data alone. An example will be given later to show that this can sometimes be done.

Let us now see these ideas work for the equations of gas dynamics (3.1.1), written in the form (3.1.2) with $\mathbf{u} = (\rho, u)$. Because the eigenvalues of A are $u \pm c$, the characteristics are the curves

$$C_+ : \frac{dx}{dt} = u + c \quad \text{and} \quad C_- : \frac{dx}{dt} = u - c. \tag{3.1.13}$$

To find the Riemann invariants we use (3.1.11). We seek eigenvectors of A^T with eigenvalues $u \pm c$; that is, vectors $(h_\pm, k_\pm)$ such that

$$\begin{bmatrix} u & c^2/\rho \\ \rho & u \end{bmatrix} \begin{bmatrix} h_\pm \\ k_\pm \end{bmatrix} = (u \pm c) \begin{bmatrix} h_\pm \\ k_\pm \end{bmatrix}.$$

This is easy to do: one finds

$$\begin{bmatrix} h_\pm \\ k_\pm \end{bmatrix} = \begin{bmatrix} \pm c/\rho \\ 1 \end{bmatrix}.$$

Thus, the Riemann invariants are found by inspection to be

$$\Gamma_\pm = u \pm \int \frac{c(\rho)}{\rho}\, d\rho, \tag{3.1.14}$$

that is,

$$\frac{\partial \Gamma_\pm}{\partial \rho} = h_\pm, \qquad \frac{\partial \Gamma_\pm}{\partial u} = k_\pm,$$

as required by (3.1.11). Thus, $\Gamma_\pm$ are constant along the $\pm$ characteristics, respectively.

The fact that Γ_+ (resp. Γ_-) is constant on C_+ (resp. C_-) does not in itself enable us to integrate the characteristic equations. The trouble is that Γ_+ need not be constant on C_-, and Γ_- need not be constant on C_+; if they were, then by inverting (3.1.14), u and ρ would be constant on characteristics and the characteristics would be straight lines. If, by some device, the characteristics can be found, then the equations can be solved as follows: From (x, t) follow the $\pm$ characteristics back to the curve on which initial data are prescribed to determine $\Gamma_\pm$ via (3.1.14). Using these values, solve (3.1.14) for (u, ρ) and the result will be the value of (u, ρ) at (x, t).

Of course, as in Example 4 there may be difficulties with two C_+ characteristics crossing. There may also be boundary conditions that have to be taken into account.

We now discuss a different method for obtaining the Riemann invariants. Consider the change of variables $u = u(x, t), \rho = \rho(x, t)$ from the (x, t) plane to the (ρ, u) plane and assume it is invertible (with nonzero Jacobian) so x and t may be regarded as functions of ρ and u. This is called the ***hodograph transformation***.

Because the transformations $(x,t) \mapsto (\rho,u)$ and $(\rho,u) \mapsto (x,t)$ are inverses, their Jacobian matrices are inverses:

$$\begin{bmatrix} \rho_x & \rho_t \\ u_x & u_t \end{bmatrix} = \begin{bmatrix} x_\rho & x_u \\ t_\rho & t_u \end{bmatrix}^{-1} = \frac{1}{J}\begin{bmatrix} t_u & -x_u \\ -t_\rho & x_\rho \end{bmatrix},$$

where $J = x_\rho t_u - x_u t_\rho \neq 0$. Substitution into (3.1.1) yields a *linear* hyperbolic system with independent variables (ρ, u) and dependent variables (x,t) as follows:

$$x_\rho - ut_\rho + \frac{c^2}{\rho} t_u = 0,$$

$$\rho t_\rho + x_u - ut_u = 0,$$

that is,

$$\begin{bmatrix} x \\ t \end{bmatrix}_\rho + \begin{bmatrix} 1 & u/\rho \\ 0 & \rho^{-1} \end{bmatrix}\begin{bmatrix} 0 & c^2/\rho \\ 1 & -u \end{bmatrix}\begin{bmatrix} x \\ t \end{bmatrix}_u = 0.$$

The eigenvalues of the coefficient matrix

$$\begin{bmatrix} 1 & u/\rho \\ 0 & \rho^{-1} \end{bmatrix}\begin{bmatrix} 0 & c^2/\rho \\ 1 & -u \end{bmatrix} = \frac{1}{\rho}\begin{bmatrix} u & c^2 - u^2 \\ 1 & -u \end{bmatrix}$$

are easily computed to be $\lambda = \pm c/\rho$. Thus, the characteristics in the (ρ, u) plane are

$$\frac{du}{d\rho} = \pm\frac{c}{\rho},$$

that is,

$$u \pm \int \frac{c(\rho)}{\rho}\, d\rho = \text{constant}.$$

Thus, we recover the fact that $\Gamma_\pm$ = constant defines the characteristics; that is, we recover the Riemann invariants obtained earlier.

As we saw earlier, it may be difficult to determine the characteristics in a general flow. However, in special circumstances, the characteristics are straight lines. We now introduce some terminology relevant to this situation.

A ***state*** of a gas is a pair of values (ρ, u). A ***constant state*** is a region in the (x,t) plane in which ρ and u are constant. A ***simple wave*** is a region in the (x,t) plane in which one of Γ_+, Γ_- is constant. If it is Γ_+ that is constant, we refer to the region as a Γ_+ ***simple wave***. A Γ_- ***simple wave*** is defined analogously.

In a Γ_+ simple wave, both Γ_- and Γ_+ are constant along C_- characteristics by definition of Γ_- . From (3.1.14), u and c are constant too, and so

C_- is a straight line. Similarly, in a Γ_- simple wave, the C_+ characteristics are straight lines.

Consider a constant state S that is bounded by a smooth curve σ in the (x, t) plane, as shown in Figure 3.1.4. (We implicitly assume that S is the largest region on which (ρ, u) are constant, that is, the region across σ exterior to S is not a constant state.) Because we are assuming $c \neq 0$, then the C_+ and C_- characteristics are not parallel at any point. Therefore, at any point P of σ, one of C_+ or C_- must cross σ. Suppose that C_+ crosses σ. Now Γ_+ is constant along each C_+ characteristic and is constant throughout S. Therefore, assuming u and ρ are smooth enough, Γ_+ will be constant on a region exterior to S. Thus *there is a portion of the* (x, t) *plane adjoining* S *that is a* Γ_+ (or Γ_-) *simple wave*. If at some point along σ, the other characteristic C_- was not equal to σ, it would cross it (again assuming smoothness) and so Γ_- would be constant on some region exterior to S. That region would thus be a constant state. Therefore, *the boundary of a constant state is composed of straight lines.*

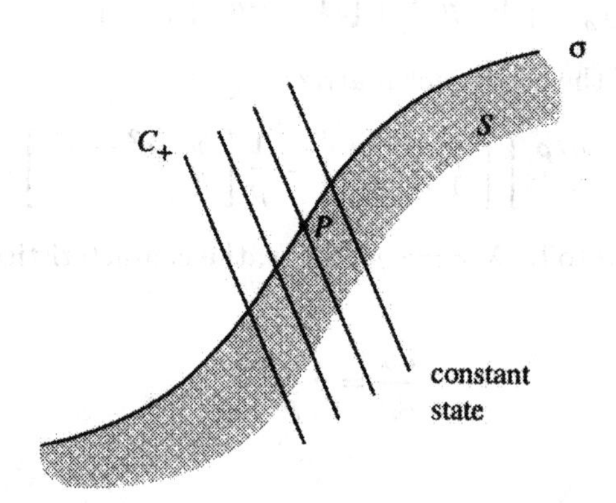

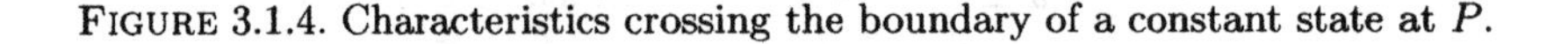

FIGURE 3.1.4. Characteristics crossing the boundary of a constant state at P.

We will now give an example that shows the usefulness of the preceding observations. Consider an infinitely long tube extending along the x-axis. We assume that gas fills the half-tube $x > 0$ with a piston situated at $x = 0$, and that the gas is in a motionless state with a constant density ρ_0 (Figure 3.1.5). At $t = 0$, we start pulling the piston to the left so that the piston follows a path $x = x(t)$ for $t > 0$ in the (x, t) plane (Figure 3.1.6).

As a result the gas is set into motion. Because the gas is originally motionless with uniform density ρ_0, there is a constant state I in the (x, t) plane where the $C_\pm$ characteristics have constant slope $\pm c(\rho_0)$. The Riemann invariants in I are

$$\Gamma_\pm^0 = \pm \int_a^{\rho_0} \frac{c(s)}{s}\, ds = \text{constant}.$$

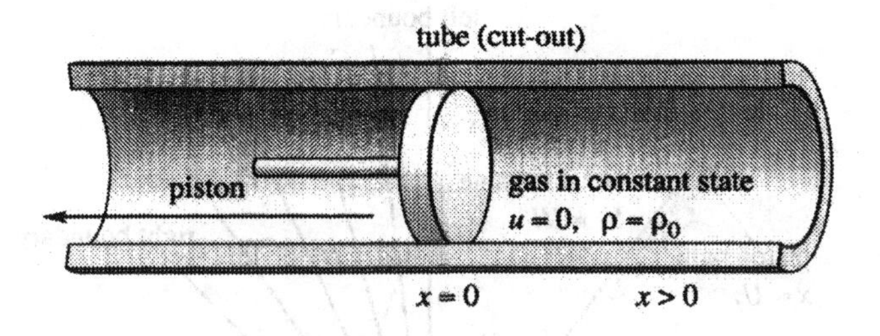

FIGURE 3.1.5. Gas in a tube with a piston at $t = 0$.

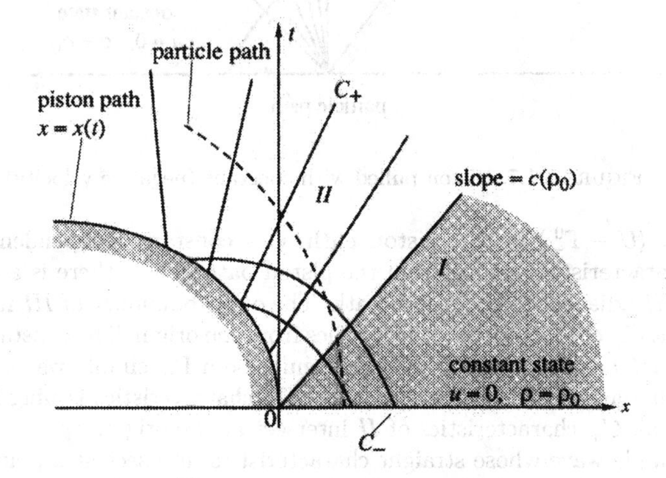

FIGURE 3.1.6. Dynamics of the gas in the piston.

The boundary of I is a straight C_+ characteristic that must emanate from the origin. This is readily seen from the fact that two distinct C_+ characteristics in a Γ_- wave cannot intersect at $t > 0$. If they did, they would coincide. The C_- characteristics of I penetrate into the adjoining region II, and they can be traced at most as far as the piston path. Hence, the region must be a Γ_- simple wave (part of it may also be a constant state). The C_+ characteristics in the region II are all straight lines. In particular, if the piston is withdrawn with a constant velocity U (which is negative), that is, $x(t) = Ut$, then we claim that *the density of the gas is constant along the piston path* (Figure 3.1.7). To prove this fact, pick any point B on the piston path. The gas particle at B moves with velocity U. Because $\Gamma_-(B) = \Gamma_-^0$, we have

$$\int_a^{\rho(B)} \frac{c(\rho)}{\rho}\, d\rho = U - \Gamma_-^0 = \text{constant}. \tag{3.1.15}$$

This is possible only if $\rho(B)$ is constant along the piston path, because $c(\rho)$ and ρ are positive. Hence, the Riemann invariant Γ_+, which is equal

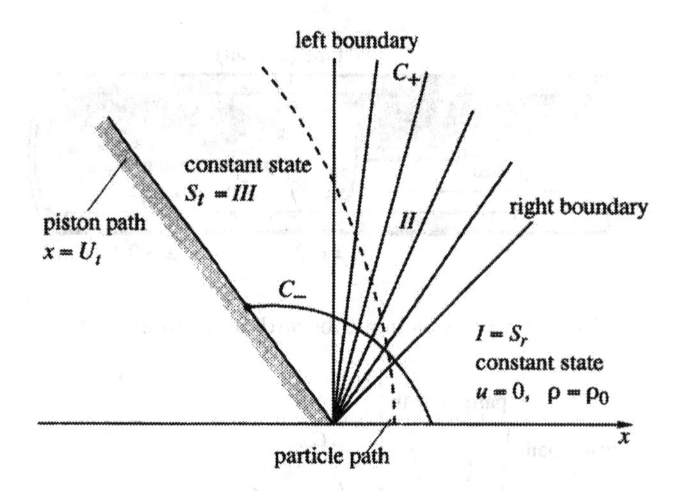

FIGURE 3.1.7. Piston pulled with constant (negative velocity).

to $U + (U - \Gamma_-^0)$ on the piston path, is a constant independent of any C_+ characteristics issuing from the piston path. Thus, there is a constant state *III* adjacent to the piston path. The other boundary of *III* must be a straight C_+ characteristic, which issues from the origin. The constant states *I* and *III* enclose a region *II* which must be a Γ_- simple wave. Because the boundary of *II* consists of straight C_+ characteristics issuing from the origin, all C_+ characteristics of *II* intersect at the origin.

A simple wave whose straight characteristics intersect at a point on the x-axis is called a ***centered rarefaction wave***. We justify this terminology by noting the following fact: From (3.1.15),

$$\int_{\rho 0}^{\rho(B)} \frac{c(\rho)}{\rho} d\rho = \int_{a}^{\rho(B)} \frac{c(\rho)}{\rho} d\rho + \Gamma_-^0 = U < 0.$$

Because $c(\rho)$ and ρ are positive, we must have $\rho(B) < \rho_0$ (i.e., in Figure 3.1.7 the gas is rarefied when the gas state passes through *II* from *I* to *III*).

We can carry out the explicit construction of the solution to the piston problem at any point (x,t) in the special case of "γ law gas" whose equation of state is given by $p = A\rho^\gamma, A, \gamma =$ constants, $\gamma > 1$.

The sound speed in such a gas is

$$c = \sqrt{\frac{dp}{d\rho}} = \sqrt{A\gamma\rho^{\gamma-1}} = \sqrt{\frac{\gamma p}{\rho}}, \tag{3.1.16}$$

and the Riemann invariants are

$$\Gamma_\pm = u \pm \int \frac{c(\rho)}{\rho} d\rho = u \pm \frac{2}{\gamma - 1} c.$$

We call the constant state I the ***right state*** S_r. In this state

$$u_r = 0, \quad \rho_r = \rho_0, \quad p_r = A\rho_0^\gamma, \quad \text{and} \quad c_r = \sqrt{\frac{\gamma p_r}{\rho_0}}.$$

We call state III the ***left state*** S_l and in it (u, ρ, p, c) have constant values, say (u_l, ρ_l, p_l, c_l). Obviously, $u_l = U$. From

$$\Gamma_- = U - \frac{2}{\gamma - 1}c_l = u_r - \frac{2}{\gamma - 1}c_r = -\frac{2}{\gamma - 1}c_r,$$

we can determine c_l and hence the entire left state from (3.1.16). Because we know c_l, the left boundary of the rarefaction wave is given by $x = (U + c_l)t$. Finally, if (x, t) is a point inside the rarefaction, then

$$u + c = \frac{x}{t}, \tag{3.1.17}$$

since the C_+ characteristic through (x, t) is a straight line issuing from the origin. On the other hand, the rarefaction is a Γ_- simple wave, and

$$\Gamma_- = u - \frac{2}{\gamma - 1}c = u_r - \frac{2}{\gamma + 1}c_r. \tag{3.1.18}$$

Relations (3.1.17) and (3.1.18) yield

$$u(x, t) = \frac{2}{\gamma + 1}\left(\frac{x}{t} - c_r\right),$$

$$c(x, t) = \left(\frac{\gamma - 1}{\gamma + 1}\right)\left(\frac{x}{t} + \frac{2}{\gamma + 1}c_r\right).$$

We now use (3.1.16) and $p = A\rho^\gamma$ to obtain $\rho(x, t)$ and $p(x, t)$. Thus we have constructed the gas flow at every point (x, t).

Two constant states, S_1 and S_2, are said to be ***connected by a*** Γ_- ***rarefaction wave*** if the configuration shown in Figure 3.1.8 is a solution of the differential equations for gas flow.

Remember that in a Γ_- simple wave, the C_+ characteristics are straight lines. The configuration in Figure 3.1.8 depicts two constant states, S_1 and S_2, connected by a Γ_- rarefaction wave. We ask the question: Given a state S_1, that is, given values ρ_1 and u_1, what are the possible states S_2 that can be connected to S_1 by means of a centered rarefaction wave? We are only interested in states S_2 whose density ρ_2 is less than ρ_1; this corresponds to the situation created when a piston moves to the left. For this discussion, we assume the relation $p = A\rho^\gamma$.

Choose p_2 such that $0 < p_2 < p_1$. We claim that p_2 will uniquely determine the state S_2 that can be connected to S_1 by a centered rarefaction wave. Given p_2, we find ρ_2 from the relation $p_2 = A\rho_2^\gamma$. Given S_1, we know ρ_1 and u_1 and thus can compute

$$\Gamma_- = u_1 - \frac{2}{\gamma - 1}c(\rho_1),$$

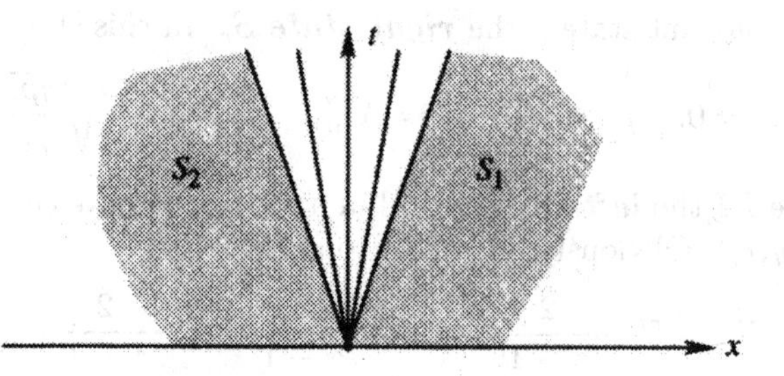

FIGURE 3.1.8. Two constant states connected by a rarefaction wave.

which is constant in S_1. Because Γ_- is constant throughout the centered rarefaction, we have

$$u_2 - \frac{2}{\gamma - 1} c(\rho_2) = u_1 - \frac{2}{\gamma - 1} c(\rho_1).$$

This determines u_2 and hence the constant state S_2.

How far can these C_+ characteristics in Figure 3.1.8 fan out? The answer is seen from the relation

$$\Gamma_- = u - \frac{2}{\gamma - 1} c(\rho);$$

that is, when ρ is zero we must stop.

The methods developed so far cannot deal with the case of a piston being pushed in. In such a case, our C_+ characteristics would soon collide, as shown in Figure 3.1.9.

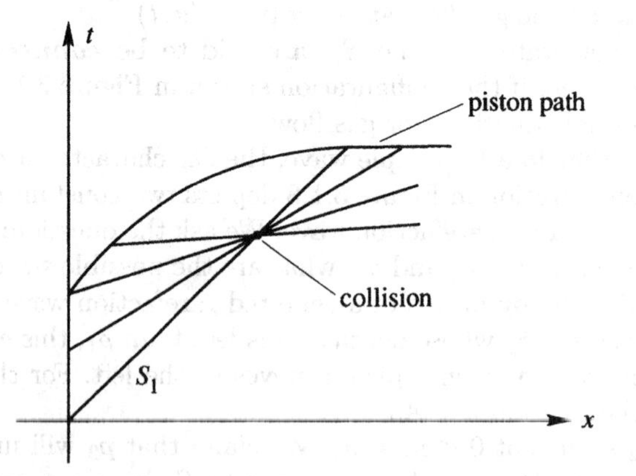

FIGURE 3.1.9. The characteristics can collide if the piston is pushed in.

When the characteristics collide, the solutions become multiple valued and thus, as presented so far, meaningless. To deal with this situation we will modify our approach and introduce the concept of a "weak solution" in the next section.

3.2 Shocks

To deal with the crossing of characteristics and the formation of shocks, we introduce some concepts from thermodynamics and the notion of a weak solution.

We begin with a brief review of relevant parts of thermodynamics for ideal flow.[1] We assume there is a specific internal energy function ϵ ("specific" means "per unit mass"). For example, as we saw in **§1.1**, for ideal gases,

$$\epsilon = \frac{p}{\rho}\left(\frac{1}{\gamma - 1}\right).$$

The total energy per unit volume is

$$e = \tfrac{1}{2}\rho u^2 + \rho\epsilon.$$

Assuming that heat does not enter the fluid domain from its boundaries, the total energy can only be reduced when the fluid does work. The work done by a fluid volume W per unit time is

$$-\int_{\partial w} p\mathbf{u} \cdot \mathbf{n}\, dA$$

where $\mathbf{n}$ is the outward unit normal to ∂W and dA is the area element. This must equal the rate of change of energy in W:

$$\partial_t \int_W e\, dV = -\int_{\partial W} p\mathbf{u} \cdot \mathbf{n}\, dA = -\int_W \operatorname{div}(p\mathbf{u})\, dV.$$

By the transport theorem in **§1.1**, this integral form is equivalent to the differential equation

$$\partial_t e + \operatorname{div}(e\mathbf{u}) + \operatorname{div}(p\mathbf{u}) = 0.$$

that is,

$$\partial_t e + \operatorname{div}((e + p)\mathbf{u}) = 0,$$

In one dimension this becomes

$$\partial_t e + ((e + p)u)_x = 0. \tag{3.2.1}$$

[1] For an "axiomatic" treatment, see Marsden and Hughes [1994]. See also L. Malvern [1969] *Introduction to the Mechanics of a Continuous Medium,* Prentice-Hall.

Let us check that (3.2.1) indeed holds for an ideal gas. Here we choose

$$p = A\rho^{\gamma} \quad \text{and} \quad \epsilon = \frac{p}{\rho}\frac{1}{\gamma - 1};$$

then (3.2.1) reads

$$\partial_t \left(\tfrac{1}{2}\rho u^2 + \frac{p}{\gamma - 1}\right) + \left(\left(\tfrac{1}{2}\rho u^2 + \frac{p}{\gamma - 1} + p\right) u\right)_x = 0.$$

Substituting $p = A\rho^{\gamma}$ and the equation of motion $u_t + uu_x = -p_x/\rho$, we see that this equation is indeed satisfied, *as long as u and ρ are smooth.*

The energy equation (3.2.1) is called the ***first law of thermodynamics***. We regard it as one of the basic equations. Our system, thus enlarged, becomes

$$\left.\begin{aligned} \rho_t + u\rho_x + \rho u_x &= 0, \\ u_t + uu_x + \frac{p_x}{\rho} &= 0, \\ e_t + ((e + p)u)_x &= 0. \end{aligned}\right\} \qquad (3.2.2)$$

When shocks form, discontinuities develop and so the meaning of equations (3.2.2) has to be carefully considered. Later we shall examine in detail why it is desirable to work with (3.2.2). For instance, for a gas (for which $e = \frac{1}{2}\rho u^2 + p/(\gamma - 1)$) we shall see that (3.2.2) can be meaningful even when $p = A\rho^{\gamma}$ breaks down. Physically, it is reasonable to concentrate on the system (3.2.2) because it expresses conservation of mass, momentum, and energy. We remark that the system (3.2.2) has no Riemann invariants in general.

The second law of thermodynamics in the form of a mathematically reasonable postulate about shock waves will be introduced later. In thermodynamics, the second law asserts the existence of an "entropy" η that has the property $D\eta/Dt \geq 0$; η increases when energy is converted from kinetic to internal energy. In works on thermodynamics, η is related to the other variables, for example, to specific heats. For an ideal gas, one can show that

$$\eta = [\,\text{constant}\,] \log(p\rho^{-\gamma}).$$

If η is increasing, we cannot have $p = A\rho^{\gamma}$, where A is a constant. When shocks form, η increases and $p = A\rho^{\gamma}$ must be abandoned. However, a suitable formulation of equations (3.2.2) will continue to make sense. We turn to this formulation next.

We now introduce the concept of a "weak solution" that will allow discontinuous solutions. The basic idea behind weak solutions is to go back to the integral formulation of the equations; this may remain valid when there is not enough differentiability to justify the differential form.[2]

Let us begin by considering general nonlinear general equations in conservation form, that is, of the form

$$u_t + (f(u))_x = 0. \tag{3.2.3}$$

Example The equation $u_t + uu_x = 0$ was discussed in the previous section. It may be written in the conservation form $u_t + \left(\frac{1}{2}u^2\right)_x = 0$. However, it can *also* be written as

$$\left(\tfrac{1}{2}u^2\right)_t + \left(\tfrac{1}{3}u^3\right)_x = 0.$$

The results one gets may depend on which conservation form is chosen. Usually, physical considerations will dictate which one is appropriate. ♦

Let $\mathbf{F} = (f(u), u)$ and note that (3.2.3) is equivalent to

$$\text{Div } \mathbf{F} = 0,$$

where $\text{Div}(f_1, f_2) = (f_1)_x + (f_2)_t$ is the ***space-time divergence***. Let φ be a smooth function with compact support in the (x, t) plane (i.e., φ vanishes outside a compact set). Then (3.2.3) is equivalent to

$$\int \varphi \cdot \text{Div}\,\mathbf{F}\, dx\, dt = 0,$$

for all such ***test functions*** φ. Integrating by parts (which is justified because φ has compact support),

$$\int \text{Grad}\,\varphi \cdot \mathbf{F}\, dx\, dt = 0. \tag{3.2.4}$$

where $\text{Grad}\,\varphi = (\varphi_x, \varphi_t)$ is the *space-time gradient* of φ. If u is smooth (3.2.3) and (3.2.4) are equivalent, but if u is not smooth, (3.2.4) can make sense even when (3.2.3) does not.

We define a ***weak solution*** of (3.2.3) to be a function u that satisfies (3.2.4) for all smooth φ with compact support.

A slightly different formulation of (3.2.4) may be used to take into account the initial conditions, say $u(x, 0) = q(x)$. Namely, we can replace the integration over all of space-time by integration over the half-space $t \geq 0$. Then during integration by parts we pick up a boundary term:

$$\int_{t\geq 0}\int_x \text{Grad}\,\varphi \cdot \mathbf{F}\, dx\, dt + \int_{-\infty}^{\infty} \varphi(x, 0)\, q\,(x)\, dx = 0. \tag{3.2.5}$$

---

[2]For a general mathematical reference, see P. D. Lax [1973] *Hyperbolic Systems of Conservation Laws and the Mathematical Theory of Shock Waves*, Conference Board of the Mathematical Sciences Regional Conference Series in Applied Mathematics, No. 11. Society for Industrial and Applied Mathematics, Philadelphia, Pa., v+48 pp.

If φ has support away from the x-axis (i.e., $\varphi = 0$ on the x-axis) then (3.2.5) reduces to (3.2.4).

So far we have the *differential form* of equation (3.2.3) and the *weak form* (3.2.4). We also have an *integral form* corresponding to the integral form of the equations of motion discussed in §**1.1**. Here it expresses "conservation of u." Indeed, consider an interval $W = [a, b]$ on the x-axis; from (3.2.3),

$$\frac{d}{dt}\int_W u\,dx = \int_W u_t\,dx = -\int_W (f(u))_x\,dx.$$

Thus,

$$\frac{d}{dt}\int_a^b u\,dx = -f(u)\Big|_a^b.$$

In particular, note that if $u \to 0$ at $\pm\infty$, then $\int_{-\infty}^{\infty} u\,dx$ is conserved in the literal sense:

$$\frac{d}{dt}\int_{-\infty}^{\infty} u\,dx = 0.$$

A weak solution does not have to be differentiable. Note, however, that a function that satisfies the integral form of the equation of motion does not have to be differentiable either, and the integral forms in fluid mechanics are in fact the basic laws we are working with, while weak solutions are merely convenient mathematical tools. A natural question to ask is: Does a weak solution necessarily satisfy the integral form of the equation? The answer is yes, and, therefore, weak solutions are in fact the objects we are looking for. (Note that the answer is, in general, affirmative only if the conserved quantities are the same in the integral forms and in the conservation forms of the equations.)

The proof of the fact that weak solutions satisfy the integral form of the equation is reasonably straightforward. Let ψ_D be the characteristic function of the domain D over which the integral form of the equation is used; *i.e,* $\psi_D(x,t) = 1$ if $(x,t) \in D$, while $\psi_D(x,t) = 0$ if (x,t) is not in D. Substitute ψ_D instead of φ into equation (3.2.4). Some straightforward manipulation of improper functions will yield the desired integral form. However, ψ_D is not an acceptable test function φ because it is not smooth. Despite this, one can find sequences of acceptable test functions $\varphi_n(x,t)$ such that

$$\int \varphi_n(x,t)\,\mathrm{div}\,\mathbf{F}\,dx\,dt \to \int \psi_D(x,t)\,\mathrm{div}\,\mathbf{F}\,dx\,dt,$$

as $n \to \infty$ for all piecewise smooth $\mathbf{F}$. During this process, one must account carefully for lines on which $\mathbf{F}$ is not smooth. We omit the details.

We now investigate properties of weak solutions of the conservation law $u_t + f(u)_x = 0$ near a jump discontinuity. Let u be a weak solution with a jump discontinuity across a smooth curve Σ in the (x, t) plane. Let φ be a smooth function vanishing outside the region S shown in Figure 3.2.1.

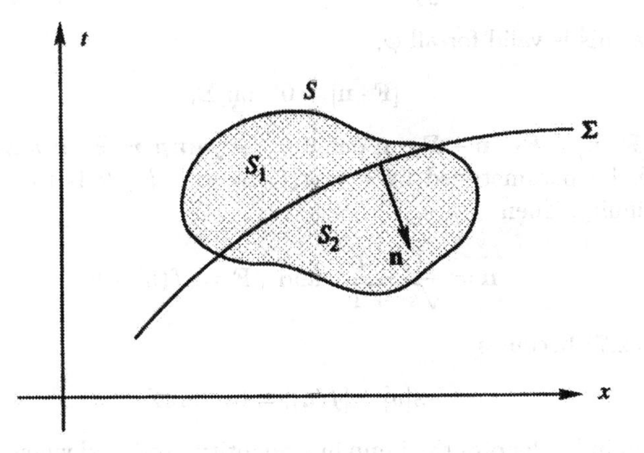

FIGURE 3.2.1. A weak solution may have a jump discontinuity across the curve Σ.

Write $S = S_1 \cup S_2$ as shown in the figure. Then

$$\begin{aligned} 0 &= \int \operatorname{Grad}\varphi \cdot \mathbf{F}\, dx\, dt \\ &= \int_{S_1} \operatorname{Grad}\varphi \cdot \mathbf{F}\, dx\, dt + \int_{S_2} \operatorname{Grad}\varphi \cdot \mathbf{F}\, dx\, dt. \end{aligned} \tag{3.2.6}$$

Let us assume that u is smooth in each of regions S_1 and S_2. Thus

$$\begin{aligned} \int_{S_1} \operatorname{grad}\varphi \cdot \mathbf{F}\, dx\, dt &= \int_{S_1} \operatorname{div}(\varphi \mathbf{F})\, dx\, dt - \int_{S_1} \varphi \operatorname{Div} \mathbf{F}\, dx\, dt \\ &= \int_{\Sigma} \varphi \mathbf{F} \cdot \mathbf{n}\, ds\, dt - \int_{S_1} \varphi \operatorname{Div} \mathbf{F}\, dx\, dt. \end{aligned}$$

In region S_1 where u is smooth, $\operatorname{Div} \mathbf{F} = 0$, and therefore

$$\int_{S_1} \operatorname{Grad}\varphi \cdot \mathbf{F}\, dx\, dt = \int_{\Sigma} \varphi \mathbf{F}_1 \cdot \mathbf{n}\, ds,$$

where in the integral over Σ, $\mathbf{F}_1$ means u is evaluated by taking the limit from region S_1. Similarly, one has

$$\int_{S_2} \operatorname{Grad}\varphi \cdot \mathbf{F}\, dx\, dt = - \int_{\Sigma} \varphi \mathbf{F}_2 \cdot \mathbf{n}\, ds.$$

The minus sign occurs because the outward normal $\mathbf{n}$ for S_1 is the inward normal for S_2. Substitution into (3.2.6) yields

$$\int_\Sigma \varphi(\mathbf{F}_1 - \mathbf{F}_2) \cdot \mathbf{n}\, ds = 0.$$

Because this is valid for all φ,

$$[\mathbf{F} \cdot \mathbf{n}] = 0 \quad \text{on } \Sigma, \tag{3.2.7}$$

where $[\mathbf{F} \cdot \mathbf{n}] = \mathbf{F}_1 \cdot \mathbf{n} - \mathbf{F}_2 \cdot \mathbf{n}$ denotes the ***jump in*** $\mathbf{F} \cdot \mathbf{n}$ ***across*** Σ.

Let Σ be parametrized by $x = x(t)$ so $s = dx/dt$ is the speed of the discontinuity. Then

$$\mathbf{n} = \frac{(1, -s)}{\sqrt{s^2 + 1}} \quad \text{and} \quad \mathbf{F} = (f(u), u);$$

thus, (3.2.7) becomes

$$-s[u] + [f(u)] = 0 \quad \text{on } \Sigma \tag{3.2.8}$$

where again [ ] denotes the jump in a quantity produced when (x, t) crosses from S_1 to S_2.

Equation (3.2.8) is the constraint that the assumed weak form of the equations (3.2.4) imposes on the values of u on both sides of a discontinuity. Also, $\int u\, dx$ is conserved if the differential form of the equation is satisfied whenever u is smooth, and (3.2.8) is satisfied across any discontinuity. It is, of course, assumed that all discontinuities are jump discontinuities. A function u that satisfies the differential equations where possible and (3.2.8) across a jump discontinuity satisfies the integral form and the weak form of the equations.

What we have just done works just as well for *systems of conservation laws*, that is, equations of the form

$$(u_i)_t + (f_i(u_1, \ldots, u_n))_x = 0, \qquad i = 1, \ldots, n$$

for a vector unknown $\mathbf{u} = (u_1, \ldots, u_n)$. Setting $\mathbf{F}_i = (f_i(u), u_i)$, these may be rewritten as

$$\operatorname{Div} \mathbf{F}_i = 0, \qquad i = 1, \ldots, n.$$

The comments on the single equation case then carry over verbatim, equation by equation, to the case of systems. In particular, each component u_i represents a conserved quantity of the system in the sense spelled out earlier.

As an example of a system of conservation laws, consider the equations of isentropic gas flow:

$$\rho_t + (\rho u)_x = 0, \qquad u_t + u u_x + \frac{p_x}{\rho} = 0.$$

They can be written in conservation form if one writes the second equation in terms of the momentum $m = \rho u$. One gets

$$\left.\begin{aligned} \rho_t + m_x &= 0, \\ m_t + \left(\frac{m^2}{\rho} + p\right)_x &= 0. \end{aligned}\right\} \tag{3.2.9}$$

In this system we assume $p = p(\rho)$. We can drop this assumption $p = p(\rho)$ if we add the energy equation

$$e_t + ((e+p)u)_x = 0,$$

where $e = \rho\epsilon + \frac{1}{2}\rho u^2$ and ϵ is given, for example, by

$$\epsilon = \frac{p}{\rho}\frac{1}{\gamma - 1}.$$

In summary, using the energy equation, we have the system of conservation laws

$$\left.\begin{aligned} \rho_t + m_x &= 0, \\ m_t + \left(\frac{m^2}{\rho} + p\right)_x &= 0, \\ e_t + \left((e+p)\frac{m}{\rho}\right)_x &= 0. \end{aligned}\right\} \tag{3.2.10}$$

The system (3.2.9) *together with* $p = A\rho^\gamma$ *will not have the same weak solutions as* (3.2.10) *together with*

$$e = \tfrac{1}{2}\rho u^2 + \rho\epsilon, \qquad \epsilon = \frac{p}{\rho}\frac{1}{\gamma - 1}.$$

It is believed on physical grounds that the condition of conservation of energy is more fundamental than $p = A\rho^\gamma$ and that, indeed, A may depend on the entropy and thus may not be constant. Therefore, we adopt the system (3.2.10) as our basic system of conservation laws. There are problems, however (e.g., in the theory of water waves), where systems such as the 2×2 system with $p = A\rho^\gamma$ are appropriate. Physical considerations dictate which quantities must be conserved.

From the jump conditions (3.2.8) applied to the system of conservation laws (3.2.10) we find that the jump relations across a discontinuity Σ in the (x, t) plane with velocity $dx/dt = s$ are given as follows:

$$\left.\begin{aligned} s[\rho] &= [m], \\ s[m] &= \left[\frac{m^2}{\rho} + p\right], \\ s[e] &= [(e+p)u]. \end{aligned}\right\} \tag{3.2.11}$$

These are called the ***Rankine–Hugoniot relations***. The first two equations in (3.2.11) are called the ***mechanical jump relations***.

The equations of fluid mechanics, like those of classical particle mechanics, are Galilean invariant. Therefore, it is legitimate to transform variables to a coordinate system moving with uniform velocity. It is convenient to pick a coordinate system whose velocity at some fixed time t_0, say $t_0 = 0$, equals that of the discontinuity and such that the discontinuity is at $x = 0$ at $t = 0$. See Figure 3.2.2.

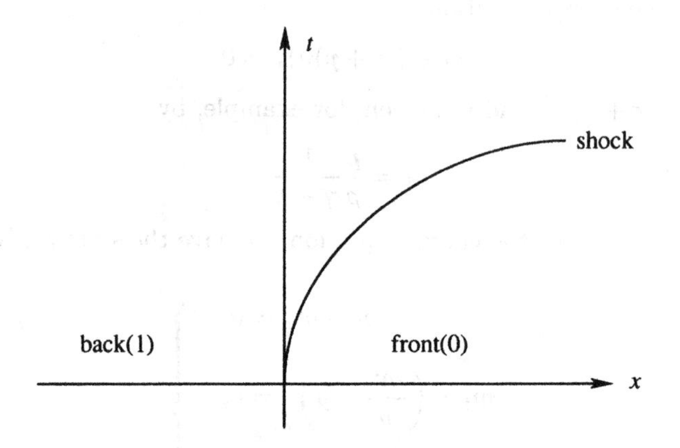

FIGURE 3.2.2. The shock in moving coordinates.

In this coordinate system, the Rankine–Hugoniot relations at the origin become

$$\left.\begin{aligned} \rho_0 u_0 &= \rho_1 u_1, \\ \rho_0 u_0^2 + p_0 &= \rho_1 u_1^2 + p_1, \\ (e_0 + p_0)u_0 &= (e_1 + p_1)u_1, \end{aligned}\right\} \tag{3.2.12}$$

where the subscripts indicate the different sides (0) or (1) of the discontinuity.

Let $M = \rho_0 u_0 = \rho_1 u_1$. If $M = 0$ we call the discontinuity a ***contact discontinuity*** or a ***slip line***. Because $u_0 = u_1 = 0$, these discontinuities move with the fluid. From $(3.2.12)_2, p_0 = p_1$, but in general $\rho_0 \neq \rho_1$.

If $M \neq 0$, we call the discontinuity a ***shock***. Because $u_0 \neq 0, u_1 \neq 0$, gas is crossing the shock, or, equivalently, the shock, is moving through the gas. The side that consists of gas that has not crossed the shock is called the ***front*** of the shock, and we identify it by a subscript 0. The other side, denoted by a subscript 1, is called the ***back***.

Some simple algebraic identities for a shock may be derived from the Rankine–Hugoniot relations. From $(3.2.12)_1$ and $(3.2.12)_2$,

$$Mu_0 + p_0 = Mu_1 + p_1, \quad i.e., \quad M = -\frac{p_0 - p_1}{u_0 - u_1}. \tag{3.2.13}$$

Substituting $u_0 = M/\rho_0 = M\tau_0$ and $u_1 = M/\rho_1 = M\tau_1$, where $\tau = 1/\rho$ is the specific volume, into (3.2.13) gives

$$M^2 = -\frac{p_0 - p_1}{\tau_0 - \tau_1}. \tag{3.2.14}$$

If we write $M^2 = M \cdot M = (\rho_0\, u_0)\,(\rho_1\, u_1)$ and $\tau = 1/\rho$, (3.2.14) becomes

$$u_0\, u_1 = \frac{p_0 - p_1}{\rho_0 - \rho_1}. \tag{3.2.15}$$

The identities (3.2.13), (3.2.14), and (3.2.15) are consequences of the mechanical jump relations only. To bring in the energy, combine $(3.2.12)_1$ with $(3.2.12)_3$ to give

$$e_0\tau_0 - e_1\,\tau_1 = p_1\,\tau_1 - p_0\,\tau_0.$$

However,

$$\begin{aligned} e_0\tau_0 - e_1\tau_1 &= \left(\tfrac{1}{2}\rho_0 u_0^2 + \rho_0\epsilon_0\right)\tau_0 - \left(\tfrac{1}{2}\rho_1 u_1^2 + \rho_1\epsilon_1\right)\tau_1 \\ &= \tfrac{1}{2}(u_0 - u_1)(u_0 + u_1) + \epsilon_0 - \epsilon_1 \\ &= -\frac{p_0 - p_1}{2M}(M\tau_0 + M\tau_1) + (\epsilon_0 - \epsilon_1) \\ &\qquad\qquad \text{(from } (3.2.12)_1 \text{ and } (3.2.13)) \\ &= (\epsilon_0 - \epsilon_1) - \frac{p_0 - p_1}{2}(\tau_0 + \tau_1). \end{aligned}$$

Thus,

$$\epsilon_1 - \epsilon_0 + \frac{p_0 + p_1}{2}(\tau_1 - \tau_0) = 0. \tag{3.2.16}$$

This relation is called the ***Hugoniot equation*** for the shock. Notice that it depends only on p and τ and not on u. Define the ***Hugoniot function*** with center (τ_0, p_0) to be

$$H(\tau, p) = \epsilon(\tau, p) - \epsilon(\tau_0, p_0) + \frac{p_0 + p}{2}(\tau - \tau_0)$$

so that (3.2.16) may be written as $H(\tau_1, p_1) = 0$.

For a "γ-law gas" for which

$$\epsilon = \frac{1}{\gamma - 1} p\tau,$$

it is easily checked that $H(\tau, p)$ is given by

$$2\mu^2 H(\tau, p) = (\tau - \mu^2\tau_0)\, p - (\tau_0 - \mu^2\tau)\, p_0,$$

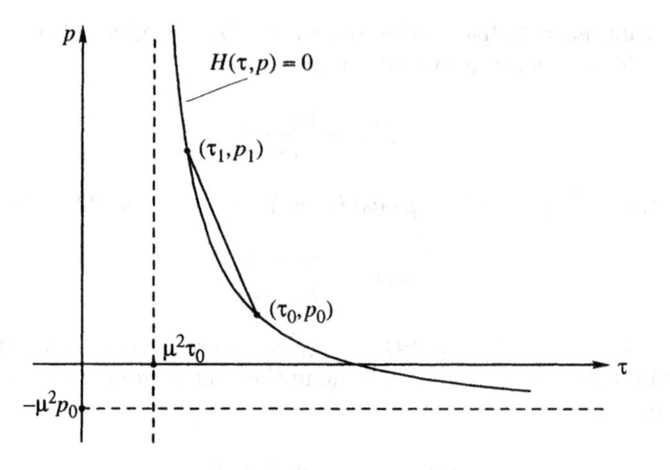

FIGURE 3.2.3. The Hugoniot curve $H(\tau, p) = 0$.

where

$$\mu^2 = \frac{\gamma - 1}{\gamma + 1}.$$

In this case, the curve $H(\tau, p) = 0$ is the hyperbola shown in Figure 3.2.3.

The Hugoniot equation states that (τ_1, p_1) lies on this hyperbola. The hyperbola represents all possible states that can be connected to the state (τ_0, p_0) through a shock. Notice from (3.2.14) that $-M^2$ is the slope of the line through (τ_1, p_1) and (τ_0, p_0) and that this then determines u_0 and u_1 via $(3.2.12)_1$. Because the curve $p\rho^{-\gamma} = p_0\rho_0^{-\gamma}$, that is, $p\tau^\gamma = p_0\tau_0^\gamma$ does not coincide with the curve $H(\tau, p) = 0$, we will not have $p\rho^{-\gamma} = \text{constant}$ across a shock, in general. Similarly, if we had insisted on (3.2.9) and $p\rho^{-\gamma} = \text{constant}$, we would have obtained shocks for which energy conservation would have been violated.

Further conditions must be imposed on weak solutions of conservation laws to select a unique, physically correct solution. The next example demonstrates nonuniqueness with $u_t + uu_x = 0$.

Example Consider the conservation law

$$u_t + \left(\frac{u^2}{2}\right)_x = 0. \tag{3.2.17}$$

Recall from §**3.1** that its characteristics are straight lines and that u is constant along them. Consider the initial data

$$u(x, 0) = \begin{cases} 0, & \text{if } x \geq 0; \\ 1, & \text{if } x < 0. \end{cases}$$

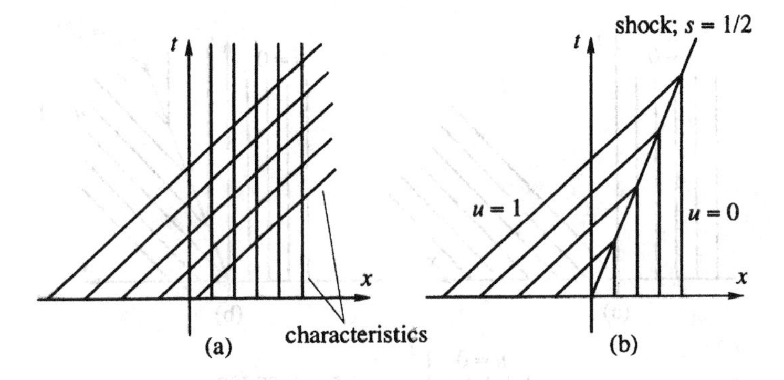

FIGURE 3.2.4. Characteristics for the initial data $u = 0$ if $x \geq 0$ and $u = 1$ if $x < 0$.

The corresponding characteristics are shown in Figure 3.2.4(a).

To keep the characteristics from crossing, we introduce a shock with propagation speed

$$s = \frac{\left[\frac{1}{2}u^2\right]}{[u]} = \frac{1}{2}$$

(see equation (3.2.8)). We thus get a globally defined weak solution by letting $u = 1$ to the left (behind) of the shock and $u = 0$ to the right (in front) of the shock. See Figure 3.2.4(b).

Next consider the initial data

$$u(x,0) = \begin{cases} 1, & \text{if } x \geq 0; \\ 0, & \text{if } x < 0. \end{cases}$$

The characteristics, shown in Figure 3.2.5(a), do not fill out the (x,t) plane.

The figure shows two ways to find a globally defined weak solution. We shall introduce a condition that excludes the solution (b) of Figure 3.2.5. ♦

Consider the characteristics of our problem and consider a shock (i.e., a discontinuity that satisfies the jump condition $-s[u] + \left[\frac{1}{2}u^2\right] = 0$) . It is easy to see that the following is true: Through every point on the path of the shock in the (x,t) plane one can draw two characteristics, one on each side of the shock; and either both of them can be traced back to the initial line or both of them can be traced upwards to the "future," that is, toward larger t. In either case, the shock is needed to avoid having characteristics intersect and create a multivalued solution. We say that this shock ***separates*** the characteristics. (We shall work out an example more general than 3.2.17 in greater detail later.) A shock is said to obey the ***entropy condition*** if the two characteristics that intersect on each point can be traced backward to the initial line, as in Figure 3.2.4(b). A shock

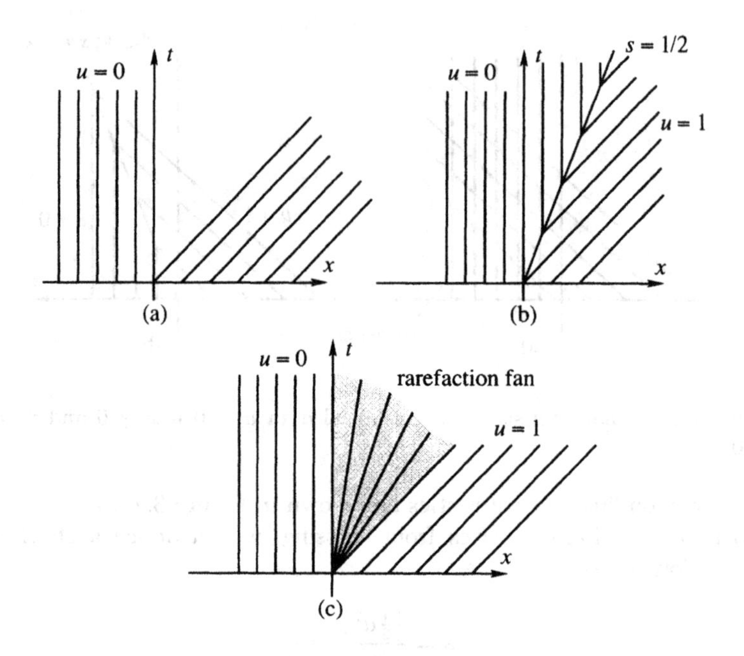

FIGURE 3.2.5. Characteristics for the initial data $u = 1$ if $x \geq 0$ and $u = 0$ if $x < 0$.

that does not obey the entropy condition is called a ***rarefaction shock***. We shall allow only shocks that do obey the entropy condition and exclude solutions such as those in Figure 3.2.5(b). This restriction will make the weak solution of the problem unique. The reason for the name "entropy condition" will appear later. The entropy condition can be viewed as a causality condition: The shock is determined by the given data, and not by future events.[3]

We shall formulate shortly the general entropy condition for shocks in systems of conservation laws. Before doing so, we list some of the reasons rarefaction shocks are excluded in gas dynamics.

1. If a rarefaction shock is allowed, then the problem will not have a unique solution.

2. A solution that includes rarefaction shocks need not depend continuously on the initial data. Specifically, in the neighborhood of a rarefaction shock, we can specify u on the left, u on the right, and s; the only required

[3]See P. D. Lax [1973] *Hyperbolic Systems of Conservation Laws and the Mathematical Theory of Shock Waves*, Conference Board of the Mathematical Sciences Regional Conference Series in Applied Mathematics, No. 11. Society for Industrial and Applied Mathematics, Philadelphia, Pa., v+48 pp.

relation is the jump condition; u is not constrained by the initial data. Thus, we can alter the solution without altering the initial data.

3. When we write the equations of gas dynamics in hyperbolic form, we neglect the viscosity. The hidden assumption is that the effect of viscosity should be small. As an example, we consider the viscous equation

$$u_t + \left(\frac{u^2}{2}\right)_x = \nu u_{xx}, \qquad \nu > 0 \tag{3.2.18}$$

corresponding to the conservation law (3.2.17). Let the solution of (3.2.18) be u_ν and let u_0 be the solution of (3.2.17) with the same initial data. Then the hidden assumption is

$$\lim_{\nu \to 0+} u_\nu = u_0.$$

In fact, one can show that $\lim_{\nu \to 0+} u_\nu$ is a weak solution of the inviscid equation (3.2.17), and one can also show that only solutions of (3.2.17) satisfying the entropy condition are the limits of the corresponding viscous solutions. We will omit the proof.[4]

4. We have seen that the quantity $\int u(x,t)\,dx$ is conserved for a solution of the conservation law (3.2.17). One can show that the "energy," that is, $\int u^2(x,t)\,dx$, cannot increase for a weak solution whose shocks satisfy the entropy condition. To make this plausible, consider a solution $u(x,t)$; we define its ***variation*** at t to be

$$\text{var}(u(x,t)) = \sup \sum_n |u(x_n,t) - u(x_{n-1},t)|$$

where the sup is taken over all the possible partitions $\{x_1, \ldots, x_N\}$ of the x-axis. (A ***partition*** is a finite division of the x-axis, $-\infty < x_1 < \cdots < x_N < \infty$.) We assume that $\text{var}(u(x,0)) < \infty$ initially. Then, because a smooth solution is propagated along characteristics from the initial data, $\text{var}(u)$ is invariant for smooth solutions (see Figure 3.2.6).

However, if $u(x,t)$ has a shock satisfying the entropy condition, $u(x,t)$ loses a part of its initial variation at any time when this shock is present. Therefore, the variation $\text{var}(u)$ cannot increase. This fact implies that $\int u^2(x,t)\,dx$ cannot increase. This is not the case with a rarefaction shock. A similar situation holds for the equations of gas dynamics.

5. In the case of the equations of gas dynamics, rarefaction shocks violate basic thermodynamic principles.

---

[4]See E. Hopf, *Comm. Pure Appl. Math.* **4** [1950], 201.

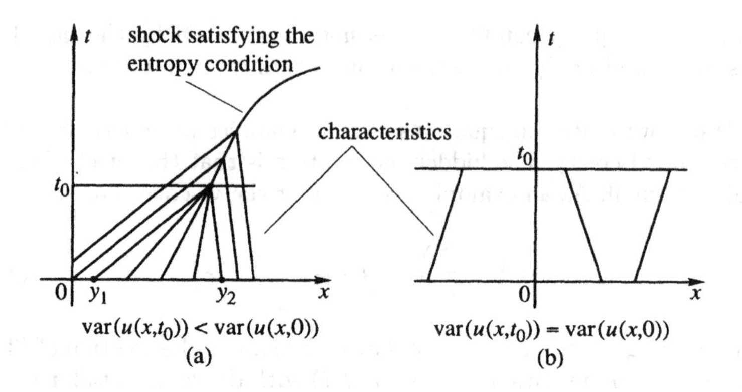

FIGURE 3.2.6. The variation is decreasing for a shock satisfying the entropy condition.

We now formulate the entropy condition for *systems* of conservation laws. Consider an n-component system $\mathbf{u}_t + (\mathbf{f}(\mathbf{u}))_x = 0$. Using the chain rule, we can write $\mathbf{u}_t + A(\mathbf{u})\mathbf{u}_x = 0$, where $A(\mathbf{u})$ is the Jacobian matrix of $\mathbf{f}$. We assume this is hyperbolic in the sense of the previous section so $A(\mathbf{u})$ has n real eigenvalues $\lambda_1, \lambda_2, \ldots, \lambda_n$. Correspondingly, we have n families of characteristics given by the curves

$$\frac{dx_i}{dt} = \lambda_i, \qquad i = 1, \ldots, n.$$

For example, in the one-dimensional case we have considered, $n = 1$ and $\lambda_1 = \lambda = u$. In the two-component case of isentropic flow, we have two families of characteristics, C_+ and C_-, associated with $\lambda_{1,2} = u \pm c$. In the 3×3 system of gas dynamics, $\lambda_1 = u + c$, $\lambda_2 = u - c$, $\lambda_3 = u$, and we have three corresponding families, C_+, C_-, and C_0, respectively.

A shock is said to ***separate characteristics of a given family*** if

i it satisfies the jump conditions;

ii through every point of the trajectory of the shock in (x,t) plane one can draw two characteristics of the family, one on each side of the shock; and

iii either both characteristics can be traced back to the initial line or they can both be traced upwards toward increasing t. (We shall see later that it is a property of the gas dynamic equations that each shock separates either the C_+ or the C_- family.)

The ***entropy condition*** for systems of conservation laws is the following: A shock satisfies the entropy condition if, when it separates characteristics of one family, the characteristics on each side can be traced back to the initial data. We shall allow only shocks that satisfy the entropy condition.

(Some authors call a discontinuity a shock only if the entropy condition is satisfied.)

A shock for gas dynamics is called ***compressive*** if the pressure behind the shock is larger than the pressure in front of the shock. Thus, in a compressive shock the pressure is raised as a fluid particle crosses the shock.

We next show that *for a γ-law gas a shock is compressive if and only if it satisfies the entropy condition.* To demonstrate this, we proceed in a number of steps.

Step 1 *The velocity of a shock is larger than the sound speed (i.e., is supersonic) on one side and is smaller than the sound speed (i.e., is subsonic) on the other.*

To see this, consider the Hugoniot relation

$$e_0\tau_0 + p_0\tau_0 = e_1\tau_1 + p_1\tau_1$$

derived in the course of establishing equation (3.2.16). We can rewrite it as

$$\tfrac{1}{2}u_0^2 + \epsilon_0 + p_0\tau_0 = \tfrac{1}{2}u_1^2 + \epsilon_1 + p_1\tau_1.$$

Substituting

$$\mu^2 = \frac{\gamma - 1}{\gamma + 1}, \quad \epsilon = \frac{p\tau}{\gamma - 1}, \quad \text{and} \quad c^2 = \gamma p \tau,$$

we get

$$\mu^2 u_0^2 + (1 - \mu^2)c_0^2 = \mu^2 u_1 + (1 - \mu^2)c_1^2. \tag{3.2.19}$$

Let c_*^2 denote the common value of both sides so that

$$(1 - \mu^2)(u_0^2 - c_0^2) = u_0^2 - c_*^2$$

and

$$(1 - \mu^2)(u_1^2 - c_1^2) = u_1^2 - c_*^2.$$

However, $\mu^2 < 1$ because $\gamma > 1$, and so

$$|u_0| > c_0 \text{ if and only if } |u_0| > c_* \tag{3.2.20}$$

and

$$|u_1| > c_1 \text{ if and only if } |u_1| > c_*$$

Rewrite the expression for c_*^2 as follows: by definition,

$$\rho_1 c_*^2 = \rho_1[\mu^2 u_1^2 + (1 - \mu^2)\gamma p_1 \tau_1] = \rho_1 \mu^2 u_1^2 + (1 - \mu^2)\gamma p_1.$$

Because $(1-\mu^2)\gamma = 1+\mu^2$,

$$\rho_1 c_*^2 = \rho_1 \mu^2 u_1^2 + (1+\mu^2) p_1.$$

From $(3.2.12)_2$ we get

$$\rho_1 c_*^2 = \mu^2 P + p_1$$

where $P = \rho_0 u_0^2 + p_0 = \rho_1 u_1^2 + p_1$. Similarly, $\rho_0 c_*^2 = \mu^2 P + p_0$. Eliminating P yields

$$c_*^2 = \frac{p_1 - p_0}{\rho_1 - \rho_0}.$$

From (3.2.15) we then get

$$c_*^2 = u_0 u_1. \tag{3.2.21}$$

This is called the ***Prandtl relation***. It follows that $|u_1| > c_*$ implies $|u_0| < c_*$, and vice versa. Together with (3.2.20), this proves our contention in Step 1.

Step 2 *Determination of the family of characteristics separated by a shock.*

Consider a shock facing to the right, that is, its front is on its right, its back is on its left, and fluid crosses it from right to left. The characteristics for our system (3.2.10) are given by

$$\frac{dx}{dt} = u + c \quad \text{(the } C_+ \text{ characteristic)}$$
$$\frac{dx}{dt} = u - c \quad \text{(the } C_- \text{ characteristic)}$$

and

$$\frac{dx}{dt} = u \qquad \text{(the } C_0 \text{ characteristic)}$$

The shock cannot separate C_0 characteristics. Indeed, by our conventions, the velocity u_0 in front of the shock is negative; by the jump relation $\rho_0 u_0 = \rho_1 u_1$, we see that u_1 is negative as well. Therefore, the configuration of the C_0 characteristics is such that on the left (labeled with a (1)) the characteristics go to the future and on the right (labeled with a (0)) they go to the past (see Figure 3.2.7), and so the shock does not separate them.

The right-facing shock cannot separate C_- characteristics either. Indeed, because $u_0 < 0$, we have $u_1 < 0$ as well and $u_0 - c < 0$, $u_1 - c < 0$; thus, one has a picture qualitatively similar to that shown in Figure 3.2.7 for C_0 characteristics.

Thus, a right-facing shock can only separate C_+ characteristics. (Similarly, a left-facing shock separates C_- characteristics.)

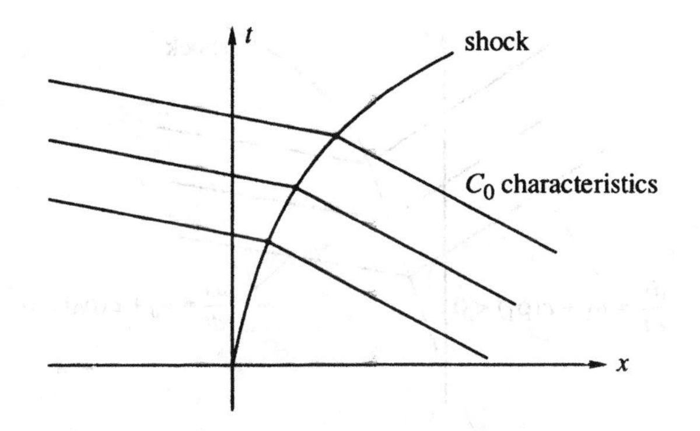

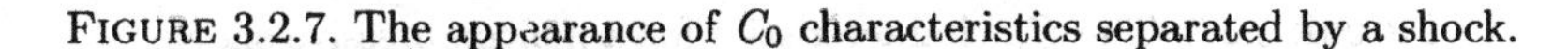
FIGURE 3.2.7. The appearance of C_0 characteristics separated by a shock.

Step 3 *A shock is compressive if and only if* $\rho_0 < \rho_1$.

Indeed, a shock is compressive when $p_1 > p_0$ by definition. From the jump condition we get $\rho_0 u_0 = \rho_1 u_1$, and thus u_1 and u_0 are both negative for a right-facing shock. From (3.2.19),

$$c_*^2 = \frac{p_1 - p_0}{\rho_1 - \rho_0} > 0;$$

therefore, $p_1 - p_0$ and $\rho_1 - \rho_0$ have the same sign, as required. A similar argument yields the same conclusion for a left-facing shock.

Step 4 *A noncompressive shock violates the entropy condition.*

A shock is noncompressive when $\rho_1 < \rho_0$, from Step 3. From the jump condition $\rho_0 u_0 = \rho_1 u_1$ we have $u_1 < u_0 < 0$. Subject to the Hugoniot constraint

$$H(\tau, p) = 0,$$

and using implicit differentiation, one sees that

$$c = \sqrt{\frac{\gamma p}{\rho}}$$

is an increasing function of ρ. From $\rho_1 < \rho_0$ and $u_1 < u_0 < 0$, we get the inequality

$$u_1 + c(\rho_1) < u_0 + c(\rho_0).$$

From Step 1, it follows that

$$|u_1| > c(\rho_1), \quad |u_0| < c(\rho_0), \quad u_1 + c(\rho_1) < 0, \quad \text{and} \quad u_0 + c(\rho_0) > 0.$$

Thus, the configuration of the slopes of the C_+ characteristics have the form shown in Figure 3.2.8. This clearly violates the entropy condition.

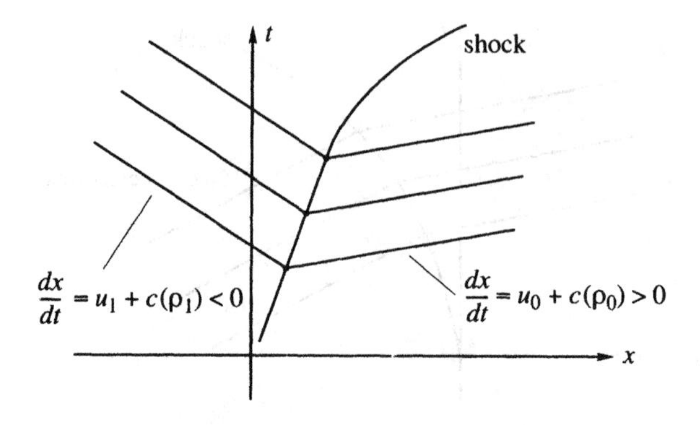

FIGURE 3.2.8. The C_+ characteristics for a noncompressive shock.

Step 5 *A compressive shock obeys the entropy condition.*

Here $\rho_1 > \rho_0$ and $u_0 < u_1 < 0$. Thus, $c(\rho_1) > c(\rho_0)$ and $u_1 + c(\rho_1) > u_0 + c(\rho_0)$. By Step 1 and Prandtl's relation we must have

$$|u_0| > c(\rho_0) \text{ and } |u_1| < c(\rho_1)$$

so that

$$u_1 + c(\rho_1) > 0 \text{ and } u_0 + c(\rho_0) < 0.$$

Thus, the characteristics have the appearance shown in Figure 3.2.9, and therefore the entropy condition is satisfied.

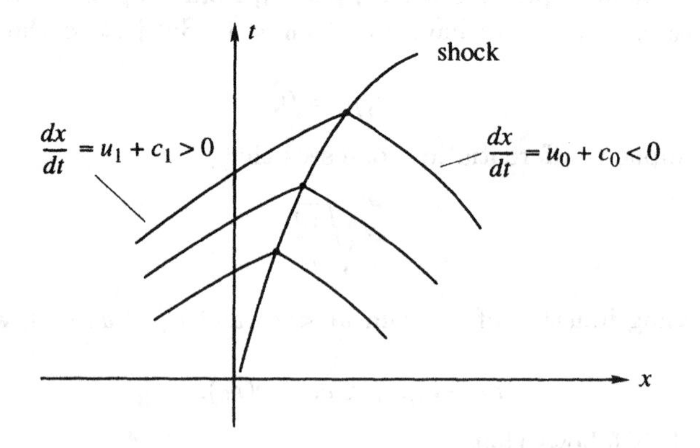

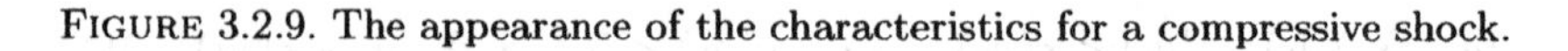
FIGURE 3.2.9. The appearance of the characteristics for a compressive shock.

We note again that if the front side were the left side, then the shock would separate the C_- characteristics, but the same conclusion would be reached.

This completes the proof of our original contention that compressive shocks for a γ–law gas obey the entropy condition.

We remarked earlier that the thermodynamic entropy can be shown to be an increasing function of $p\rho^{-\gamma}$ and that the entropy should increase across a shock. This is easy to see by consulting Figure 3.2.10.

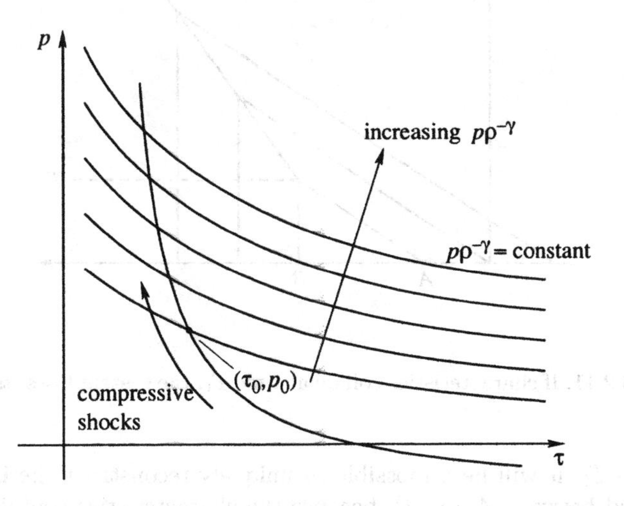

FIGURE 3.2.10. Entropy is an increasing function of $p\rho^{-\gamma}$ and increases across a shock.

Indeed, the curve $H(\tau, p) = 0$ through (τ_0, p_0) is drawn as well as the curves $p\rho^{-\gamma} = \text{constant}$; one sees that $p\rho^{-\gamma}$ increases as we move along $H(\tau, p) = 0$ in the direction of increasing p.

Thus, for a γ-law gas, we see that the geometric entropy condition we have seen so far and the condition on entropy that one can obtain from thermodynamics, happen to coincide. We shall see in §**3.4** that this is not always the case. The geometric condition is a stronger condition and may be needed for the construction of a unique solution even under circumstances where thermodynamics has little to say. We already saw such an example in the equation $u_t + uu_x = 0$.

We close this section with two remarks.

Remark 1 As long as the solution of our system of equations is smooth, the initial value problem is reversible. For example consider the conservation law $u_t + (u^2/2)_x = 0$. If the solution is smooth for all t satisfying $0 \leq t \leq T$, and if we know the solution $u(x, t)$, then by the change of variables $u = -u, t = -t$ we may solve backward to recover the initial data $u(x, 0)$.

As soon as our solution becomes discontinuous, then the reversibility of the solution is lost. Consider a solution with prescribed initial data $u(x, 0)$, where the characteristics first collide at time T_1; see Figure 3.2.11. At a later

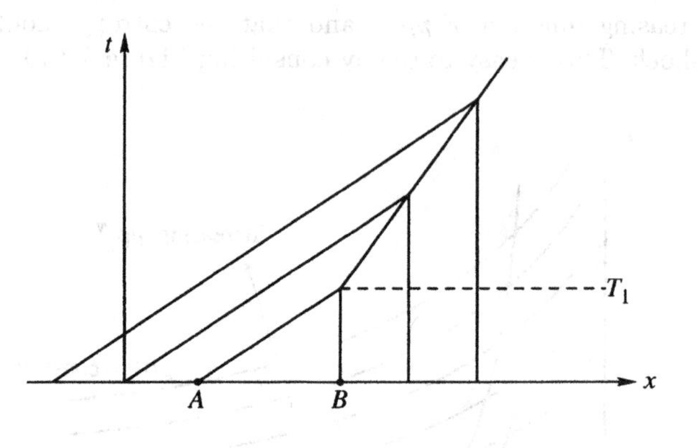

FIGURE 3.2.11. If characteristics collide at time T_1, the system loses its reversibility.

time $T > T_1$, it will be impossible to uniquely reconstruct the initial data prescribed between A and B, because the characteristics containing these initial data have been "swallowed up" by the shock. Thus, the information between A and B is "lost," which confirms the notion that an increase in entropy means that information has been lost. ♦

Remark 2 In the case of nonconstant $p\rho^{-\gamma}$, we say that a ***state*** of a gas is a set of three values $S = (\rho, u, p)$. We ask the question: What constant states can be connected to a given state by a shock? (see Figure 3.2.12.) ♦

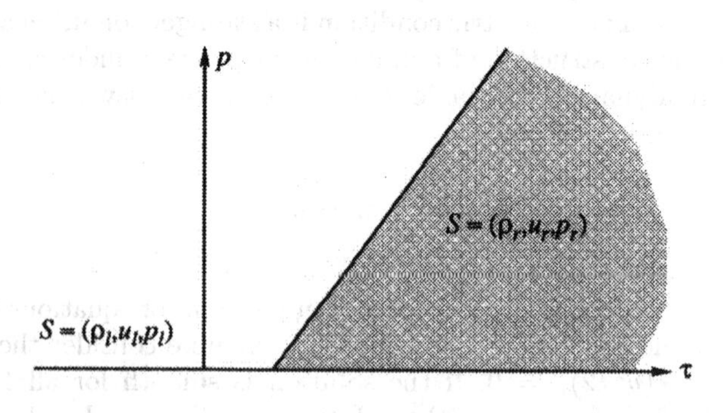

FIGURE 3.2.12. Can the left and right states be connected by a shock?

The Hugoniot relation yields a curve on which lie states that can be connected to $S_r = (\rho_r, u_r, p_r)$ by a shock; the curve represents the set of possible transitions. Furthermore, *a shock that separates two constant states must be a straight line in the (x,t) plane*, because both $[f(u)]$ and $[u]$ are constant in time and $s[u] + [f(u)] = 0$, thus $s =$ constant. Define a ***centered wave*** to be either a straight line shock or an isentropic centered rarefaction.

Then, given a right state S_r, and a pressure $p_l \geq 0$ in a left state S_l, we can find p_l , u_l such that S_l is connected to S_r by a centered wave. If $p_l > p_r$, we can find a straight line shock connecting the two states, and if $p_l \leq p_r$, we can find a centered rarefaction wave that connects them.

3.3 The Riemann Problem

The conservation laws for a γ-law gas were shown in the previous section to be

$$\begin{aligned} \rho_t + (\rho u)_x &= 0, \\ (\rho u)_t + (\rho u^2 + p)_x &= 0, \\ e_t + ((e+p)u)_x &= 0, \end{aligned} \tag{3.3.1}$$

where $e = \frac{1}{2}\rho u^2 + p/(\gamma - 1)$. The ***state*** of the gas is a vector

$$\mathbf{u} = \begin{bmatrix} \rho \\ u \\ e \end{bmatrix}.$$

The ***Riemann problem*** is the initial value problem for (3.3.1) with special initial data of the form

$$\mathbf{u}(x,0) = \begin{cases} \mathbf{u}_r, & x \geq 0, \\ \mathbf{u}_l, & x \leq 0, \end{cases}$$

where

$$\mathbf{u}_r = \begin{bmatrix} \rho_r \\ u_r \\ e_r \end{bmatrix} \quad \text{and} \quad \mathbf{u}_l = \begin{bmatrix} \rho_l \\ u_l \\ e_l \end{bmatrix}$$

are two constant states of the gas. The special case in which the velocity components vanish, that is, $u_r = 0$ and $u_l = 0$, is called a ***shock tube problem***.

The main objective of this section is to show how to solve the Riemann problem and how to use the solution to construct the solution of (3.3.1) with *general* initial data.

Consider a change of variables

$$x' = Lx, \quad t' = Lt,$$

where $L > 0$. Clearly this leaves the form of equations (3.3.1) unchanged and the initial data are unchanged as well. Thus, if we assume that the solution is unique, then

$$\mathbf{u}(x,t) = \mathbf{u}(x',t') = \mathbf{u}(xL,tL) = \mathbf{u}\left(\frac{x}{t}\right), \qquad t > 0.$$

Thus, the solution of the Riemann problem is constant along straight lines issuing from the origin in the (x,t) plane. Because hyperbolic equations have a finite speed of propagation of data, we conclude that at any instant, the state of the gas is $\mathbf{u}_r$ far enough to the right and is $\mathbf{u}_l$ far enough to the left. Let the corresponding regions in the (x,t) plane be denoted S_r and S_l, respectively. From the results of §**3.1**, the boundary of S_r is either a C_+ characteristic of (3.3.1) or a jump discontinuity emanating from the origin. A similar statement holds for S_l .

We will now give a plausible reason why S_r *can be connected to* S_l *through a right-centered wave, a constant state I, a slip line, a constant state II, and a left-centered wave* (see Figure 3.3.1). Recall from the end of the previous section that a ***centered wave*** is either a centered isentropic rarefaction wave or a shock. We now sharpen our definitions slightly. A ***right-centered wave*** is a wave that is either a shock wave facing to the right or an isentropic centered rarefaction wave in which Γ_- is constant.

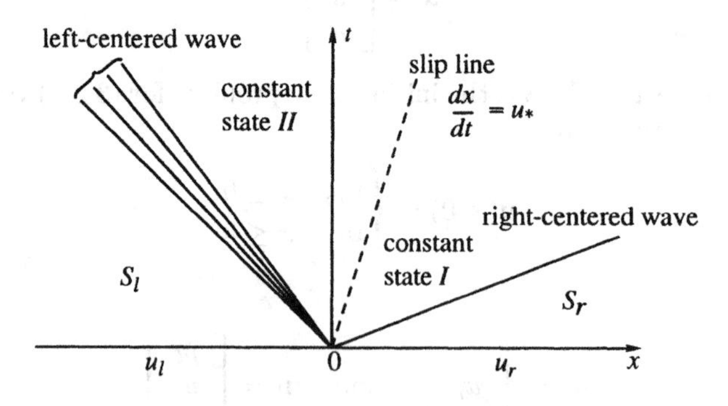

FIGURE 3.3.1. The constant states S_r and S_l connected through constant states, a slip line, and a left-centered wave.

Similarly, a ***left-centered wave*** is a wave that is either a shock wave facing to the left or a Γ_+ isentropic rarefaction. From the discussion at the end of the previous section, given a right state S_r, we can find a one-parameter family of constant states I (parametrized by the pressure, for example), which can be connected to S_r by a right-centered wave. Note that because S_r is a constant state, $p\rho^{-\gamma}$ is a constant in S_r, and if the right wave is a rarefaction, $p\rho^{-\gamma}$ is also constant in that rarefaction; thus, it is consistent to allow only isentropic rarefaction waves in the definition of a right-centered wave. Because the density is the only quantity that can be discontinuous across a slip line, we have a two-parameter family of constant states II that can be connected through a slip line, a constant state I, and a right-centered wave to S_r. Continuing once more, we get a three-parameter family of states that can be connected to S_r by a left-centered wave, a constant state II, a slip line, a constant state I, and a right-centered wave. The question is whether or not we can choose the parameters so that we end up with the desired constant state $\mathbf{u}_l$. If we can, we have a solution of the Riemann problem. If $\mathbf{u}_l$ and $\mathbf{u}_r$ are close enough, one can demonstrate the result by means of the implicit function theorem. To demonstrate the result for general $\mathbf{u}_l$ and $\mathbf{u}_r$ (for a γ-law gas), let p_* and u_* denote the pressure and velocity of the constant states I and II, respectively. Define the quantities

$$M_r = -\frac{p_r - p_*}{u_r - u_*}, \quad M_l = -\frac{p_l - p_*}{u_l - u_*}$$

(see formula (3.2.13)). We claim that these quantities are functions of p_* only, that is, $M_r = M_r(p_*)$ and $M_l = M_l(p_*)$. To prove this claim, we assume first that the right wave is a shock. Then $p_* > p_r$. From (3.2.14),

$$M_r^2 = -\frac{p_r - p_*}{\tau_r - \tau_l} \tag{3.3.2}$$

where τ_l is the specific volume of the constant state I. The Hugoniot equation (3.2.16) gives

$$(\tau_l - \mu^2\tau_r)p_* = (\tau_r - \mu^2\tau_l)p_r$$

where $\mu^2 = (\gamma - 1)/(\gamma + 1)$. Therefore,

$$\tau_l = \tau_r \frac{p_r + \mu^2 p_*}{\mu^2 p_r + p_*}.$$

Thus, using (3.3.2),

$$M_r^2 = \rho_r \left(\frac{\mu^2}{1 - \mu^2} p_r + \frac{1}{1 - \mu^2} p_* \right).$$

This shows that M_r is a function of p_* only, because the state S_r is known from the initial data. If the right wave is not a shock, it is a centered rarefaction wave. Of course, the same argument applies to M_l. On the other hand, from the definition of M_r and M_l,

$$u_* = u_*(M_r) = u_r + \frac{p_r - p_*}{M_r},$$

$$u_* = u_*(M_l) = u_l + \frac{p_l - p_*}{M_l},$$

Because u_* is continuous across the slip line, we must have

$$u_*(M_r) = u_*(M_l).$$

This is an equation for p_*. Some elementary algebra shows that this equation *has a unique real solution p_* (for a γ-law gas)*. Hence, the Riemann problem can be solved. In particular, in the case of the shock tube problem, one of the waves must be a shock and the other a rarefaction wave. The position of the slip line determines which one is the shock. If the slip line moves with positive velocity, that is, in the positive x-direction, then it acts as if it were a piston pushing into the gas in the positive x-axis. Therefore, the right wave must be a shock, and the left wave is a rarefaction.[5]

Suppose now that we have arbitrary initial data $\mathbf{u}(x,0) = \mathbf{f}(x)$ for the equation (3.3.1) where we assume $\mathbf{f}$ to be of compact support for convenience. We will use the preceding solution of the Riemann problem to construct a family of approximate solutions that converge to a solution of our initial value problem for (3.3.1). The x-axis and the t-axis are divided into intervals of length h and k, respectively. The approximate solution is to be computed at the mesh points (ih, jk) and $\left(\left(i+\frac{1}{2}\right)h, \left(j+\frac{1}{2}\right)k\right)$. We let

$$\mathbf{u}_i^j \quad \text{and} \quad \mathbf{u}_{i+1/2}^{j+1/2}$$

approximate

$$\mathbf{u}(ih, jk) \quad \text{and} \quad \mathbf{u}\left(\left(i+\tfrac{1}{2}\right)h, \left(j+\tfrac{1}{2}\right)k\right),$$

respectively. Thus, initially,

$$\mathbf{u}_i^0 = \mathbf{f}(ih), \qquad |i| = 0, 1, 2, \ldots.$$

(See Figure 3.3.2.)

The method described next permits us to advance by a time step k. Suppose we have already constructed the values of $\mathbf{u}_i^j$ at time $t = jk$ for each i. To compute $\mathbf{u}_{i+1/2}^{j+1/2}$, we consider the Riemann problem for (3.3.1) with the initial data

$$\mathbf{u}(x,0) = \begin{cases} \mathbf{u}_i^j, & x < \left(i+\frac{1}{2}\right)h; \\ \mathbf{u}_{i+1}^j, & x \geq \left(i+\frac{1}{2}\right)h. \end{cases}$$

Let $\mathbf{v}(x,t)$ be the solution to this Riemann problem. We will define

$$\mathbf{u}_{i+1/2}^{j+1/2}$$

[5]This solution is given in S. K. Godunov, *Mat. Sbornik* **47** [1957], 537. See also P. D. Lax, *Comm. Pure Appl. Math.* **10** [1957], 537.

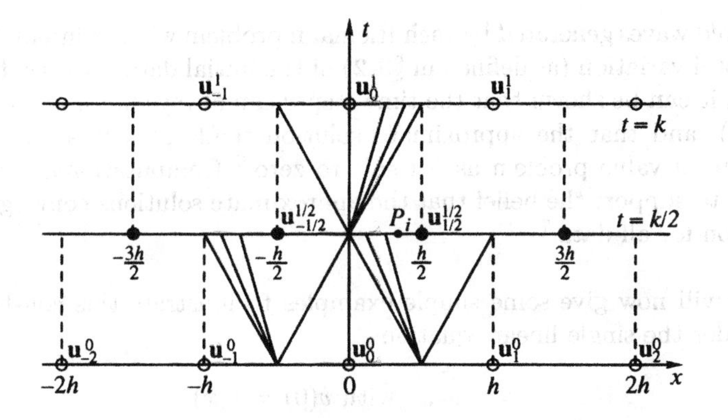

FIGURE 3.3.2. Construction of the solution of (3.3.1) for general initial data.

to be a randomly chosen value of $\mathbf{v}(x, k/2)$ on the interval $ih \leq x \leq (i+1)h$. For this purpose, define

$$\Lambda^j = \max_i \left(\left| u_i^j \right| + c_i^j \right)$$

where u_i^j is the velocity component of $\mathbf{u}_i^j$ and c_i^j is the corresponding sound speed obtained from the formula $c = \sqrt{\gamma p/\rho}$.

We require that the time step k satisfy the ***Courant–Friedrichs–Lewy condition***

$$\Lambda^j k \leq h \quad \text{for all } j. \tag{3.3.3}$$

We pick a random variable θ equidistributed in $[-1/2, 1/2]$, that is, θ has the probability density function that takes the value 1 in $[-1/2, 1/2]$, and is zero otherwise. Then we define

$$\mathbf{u}_{i+1/2}^{j+1/2} = \mathbf{v}(P_i),$$

where $P_i = ((i + 1/2)\, h + \theta h, j + k/2)$. Therefore, we have obtained the approximate values

$$\mathbf{u}_{i+1/2}^{j+1/2} \quad \text{at time } t = \left(j + \tfrac{1}{2}\right) k.$$

With exactly the same construction (which involves another independent random variable θ equidistributed in $[-1/2, 1/2]$), we can advance a further half-time step to get the approximate values

$$\mathbf{u}_i^{j+1} \quad \text{at time } t = (j + 1)k.$$

Thus, we have constructed an approximate solution. We note that condition (3.3.3) is necessary to ensure that, at each half-time step construction, the centered waves generated by each Riemann problem will not interact. When the total variation (as defined in §**3.2**) of the initial data $\mathbf{f}(x)$ is sufficiently small, it can be shown that the time step k can always be chosen to satisfy (3.3.3), and that the approximate solution tends to a weak solution of our initial value problem as h tends to zero.[6] Computational experience seems to support the belief that the approximate solutions converge to the solution for all data.[7]

We will now give some simple examples to illustrate this construction. Consider the single linear equation

$$v_t = v_x, \quad \text{with } v(0) = f(x).$$

We already know that the solution is $v(x,t) = f(x+t)$. The initial data propagate along the characteristics with speed 1. Thus, $\Lambda^j = 1$ in each step of our construction, and the Courant–Friedrichs–Lewy condition (3.3.3) becomes

$$k \leq h.$$

We will carry out the construction under this condition. To compute

$$\mathbf{v}_{i+1/2}^{j+1/2} \quad \text{from } \mathbf{v}_i^j \text{ and } \mathbf{v}_{i+1}^j,$$

we consider the Riemann problem

$$v(x,0) = \mathbf{v}_i^j \qquad \text{for } x < \left(i + \tfrac{1}{2}\right) h,$$

and

$$v(x,0) = \mathbf{v}_i^{j+1} \quad \text{for } x \geq \left(i + \tfrac{1}{2}\right) h,$$

The initial discontinuity propagates along the characteristic passing through the point $\left(\left(i+\frac{1}{2}\right)h, 0\right)$, as in Figure 3.3.3.

Hence, if the randomly chosen point P_i lies to the right of the characteristic, then

$$\mathbf{v}_{i+1/2}^{j+1/2} = \mathbf{v}_{i+1}^j,$$

that is, the solution moves $h/2$ to the left. Because this characteristic intersects the line $t = k/2$ at the point

$$\left(\left(i + \frac{1}{2}\right)h - \frac{k}{2}, \frac{k}{2}\right)$$

---

[6] See J. Glimm, "Solution in the large of hyperbolic conservation laws," *Comm. Pure Appl. Math.* **18** [1965], 69.

[7] A. J. Chorin, "Random choice solution of hyperbolic systems," *J. Comp. Phys.* **22** [1976], 519.

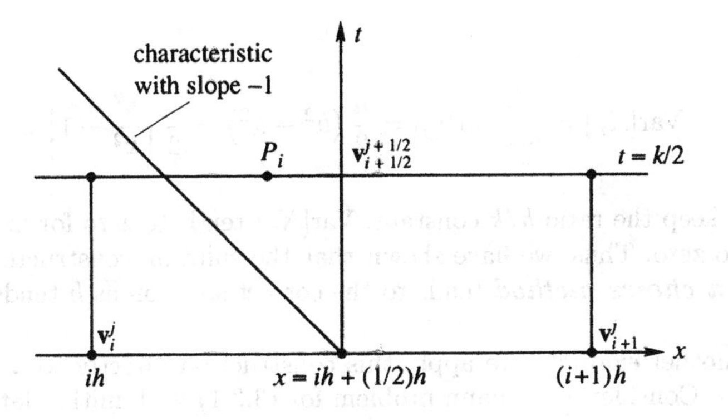

FIGURE 3.3.3. The Riemann problem for $v_t = v_x$.

and θ is uniformly distributed, P_i lies to the right of the characteristic with probability $(h+k)/(2h)$ and lies to the left with probability $(h-k)/(2h)$. Therefore, after $2n$ steps, the approximate value v_i^n at $x = ih$ and $t = nk$ comes from the initial disturbance f at the position $x - X_n$, where

$$X_n = \sum_{j=1}^{2n} \eta_j,$$

and η_j are independent, identically distributed random variables with the probability distribution

$$\text{Prob}\left[\eta_j = -\frac{h}{2}\right] = \frac{1}{2h}(h+k),$$

and

$$\text{Prob}\left[\eta_j = \frac{h}{2}\right] = \frac{1}{2h}(h-k).$$

The expectation and the variance of $\eta = \eta_j$ are

$$E[\eta] = \frac{1}{2h}(h+k)\left[-\frac{h}{2}\right] + \frac{1}{2h}(h-k)\frac{h}{2} = -\frac{k}{2},$$

and

$$\text{Var}[\eta] = E[\eta^2] - (E[\eta])^2 = \tfrac{1}{4}\left(h^2 - k^2\right).$$

Hence,

$$E[X_n] = \sum_{j=1}^{2n} E[\eta_j] = -nk = -t,$$

and

$$\text{Var}[X_n] = \sum_{j=1}^{2n} \text{Var}[\eta_j] = \frac{n}{2}\left(h^2 - k^2\right) = \frac{k}{2}\left[\frac{h^2}{k^2} - 1\right] t.$$

If we keep the ratio h/k constant, $\text{Var}[X_n]$ tends to zero for fixed t as h tends to zero. Thus, we have shown that the solution constructed by this ***random choice method*** tends to the correct solution as h tends to zero.

As another example, we apply this construction directly to a Riemann problem. Consider a Riemann problem for (3.3.1) with initial data

$$\mathbf{u}(x, 0) = \begin{cases} \mathbf{u}_r & \text{for } x \geq 0, \\ \mathbf{u}_l & \text{for } x < 0. \end{cases}$$

We assume that corresponding regions S_r and S_l can be connected by a shock propagating with speed s. When we divide the x-axis into pieces of length h, we have only one nontrivial Riemann problem at each half-time step construction (see Figure 3.3.4).

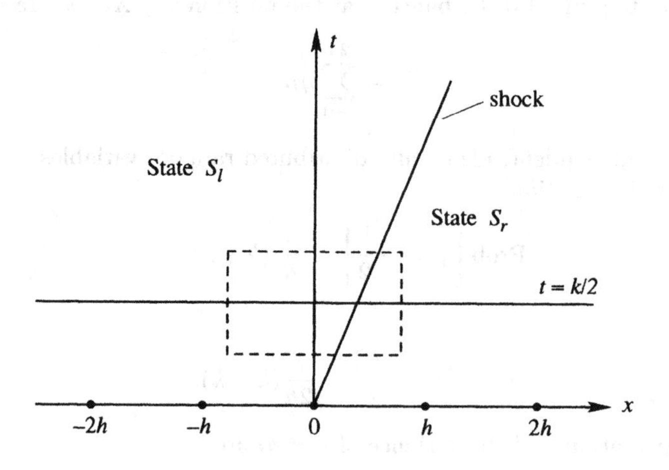

FIGURE 3.3.4. Breaking a Riemann problem into smaller ones.

This nontrivial Riemann problem is also solved by connecting S_r to S_l with the *appropriate* connecting shock. Hence, the argument given for a single linear equation shows that the discontinuity of the approximate solution is at a position whose expected value is correct, and that the variance of the position also tends to zero as h tends to zero, that is, the approximate solution tends to the correct solution. A similar conclusion applies to the case where S_r can be connected to S_l either by a centered rarefaction wave or by a slip line. Thus, we obtain the conclusion that, if we apply this random choice method directly to the simplest Riemann problems, the approximate solutions tend to the correct solution as h tends to zero. Note that even if the flow is not isentropic, that is, $p\rho^{-\gamma} \neq$ constant, the construction uses Riemann problems in which $p\rho^{-\gamma} =$ constant on either side of the slip line in the absence of a shock.

The construction just presented is the basis of Glimm's existence proof for hyperbolic systems.[8] For the 2×2 isentropic flow equations, other methods of proof are available.[9]

A major component in the Glimm construction is the solution of the Riemann problem. We have offered a construction of a solution of the Riemann problem but have said nothing about its uniqueness. The existence and uniqueness of the solution of the Riemann problem for gas dynamics subject to an appropriate formulation of the entropy condition have been established.[10]

It is worth noting that not very much is known rigorously about the solutions of the equation of gas dynamics in more than one space variable. (See Majda's book listed in the Preface for some information.)

Glimm's construction is also the basis for numerical algorithms.[11] The strategy for picking the numbers θ is the determining factor in the accuracy of these algorithms.

3.4 Combustion Waves

This section examines some additional features of conservation laws and applies them to a modified system of gas dynamic equations that allows for combustion. We shall begin by studying a single conservation law

$$u_t + (f(u))_x = 0 \tag{3.4.1}$$

and concentrate on the information the shape of the graph of f gives about discontinuities. Later, we generalize the results to systems and apply them to our specific system. We examine in four steps the cases where the graph of f is straight, concave up, concave down, and, finally, neither concave up nor down.

Case 1 ***f is linear***; that is, $f(u) = au$, $a =$ constant.

---

[8] J. Glimm, Loc. cit.

[9] See, for example, R. DiPerna, "Convergence of the Viscosity Method for Isentropic Gas Dynamics," *Comm. Math. Phys.*, **91** [1983], 1.

[10] See T. P. Liu, "The Riemann Problem for General Systems of Conservation Laws," *J. Differential Equations*, **18** [1975], 218.

[11] See A. J. Chorin, "Random Choice Solution of Hyperbolic Systems," *J. Comp. Phys.*, **22** [1976], 517.

Here (3.4.1) becomes

$$u_t + au_x = 0 \tag{3.4.2}$$

whose characteristics are the straight lines $dx/dt = a$. Because the characteristics do not intersect, smooth initial data propagate to a smooth solution; in fact, the solution with $u(x,0) = \varphi(x)$ is

$$u(x,t) = \varphi(x - at).$$

However, discontinuities in the initial data are propagated along characteristics. This result is also a consequence of our formula for the speed of the discontinuity (see equation (3.2.8))

$$s = \frac{f(u_l) - f(u_r)}{u_l - u_r} = a,$$

where u_l is the value of u to the left of the discontinuity and u_r that to the right. In particular, a discontinuity must be a characteristic. See Figure 3.4.1.

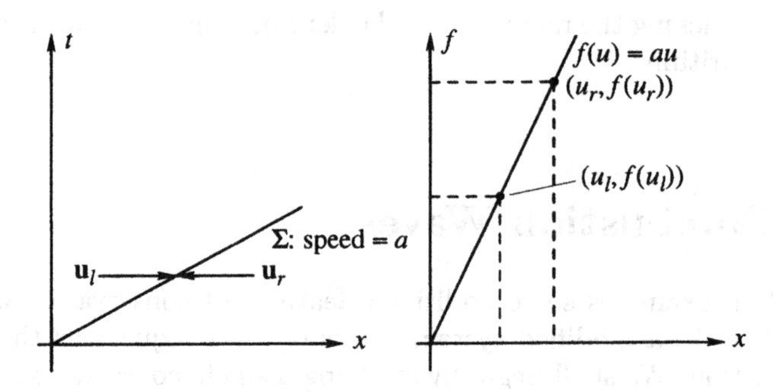

FIGURE 3.4.1. If f is linear, discontinuities move along characteristics.

Case 2 ***f is concave up***; that is, $f_{uu} > 0$. (An example is $u_t + uu_x = 0$, whose characteristics are again straight lines, as we showed in §**3.2**, and where $f(u) = u^2/2$.)

The characteristics of (3.4.2) are given by

$$\frac{dx}{dt} = f'(u)$$

because (3.4.1) is equivalent to

$$u_t + f'(u)u_x = 0$$

in the region where u is smooth. We still have

$$s = \frac{f(u_l) - f(u_r)}{u_l - u_r}$$

which, by the mean value theorem, gives $s = f'(\xi)$ for some ξ between u_l and u_r. See Figure 3.4.2.

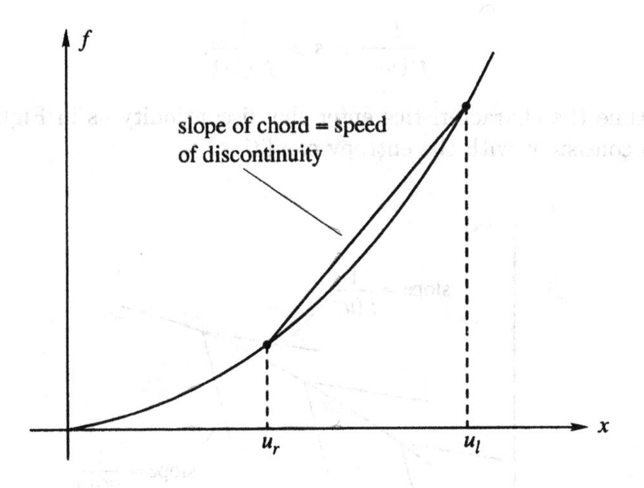

FIGURE 3.4.2. The discontinuity speed in the case f is concave up.

Let us examine the two possibilities $u_r < u_l$ and $u_l < u_r$ separately. Because $f_{uu} > 0$, f' is increasing. Thus, if $u_l < u_r$, $f'(u_l) < s < f'(u_r)$ and

$$\frac{1}{f'(u_l)} > \frac{1}{s} > \frac{1}{f'(u_r)}.$$

Thus, the slopes of the characteristics have the configuration shown in Figure 3.4.3.

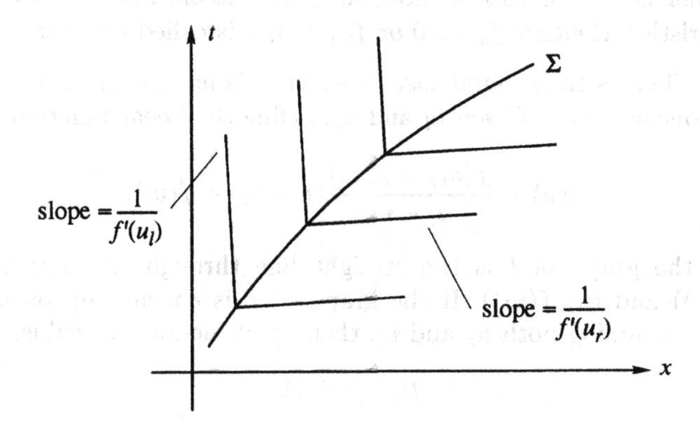

FIGURE 3.4.3. The slopes of the characteristics for f concave up and $u_l < u_r$.

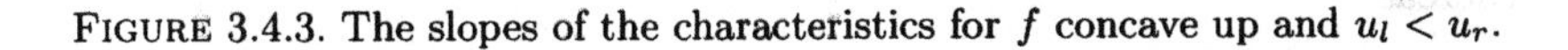

We get a shock configuration that violates the entropy condition. Thus, if $u_l < u_r$ these states should not be connected by a shock; a rarefaction fan must be used.

On the other hand, if $u_r < u_l$, then $f'(u_r) < s < f'(u_l)$ and

$$\frac{1}{f'(u_r)} > s > \frac{1}{f'(u_l)}.$$

This time the characteristics enter the discontinuity as in Figure 3.4.4, which is consistent with the entropy condition.

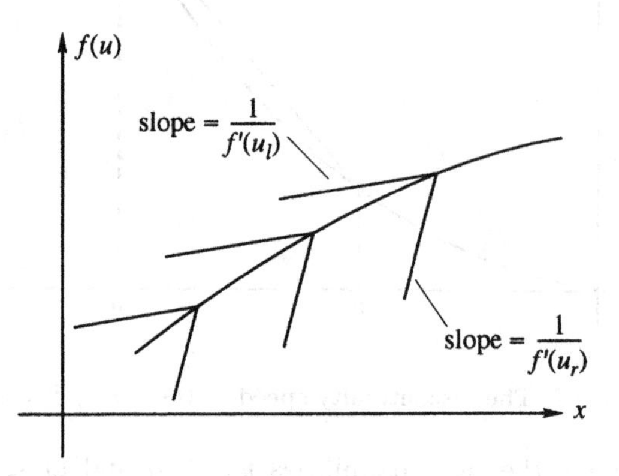

FIGURE 3.4.4. The slopes of the characteristics for f concave up and $u_r < u_l$.

Case 3 ***f is concave down***, that is, $f_{uu} < 0$. An argument similar to that of case 2 shows that for a shock we must have $u_l < u_r$ and for a rarefaction fan, $u_r < u_l$.

In either case 2 or case 3, note that the discontinuity ***separates*** the characteristics. If either $f_{uu} < 0$ or $f_{uu} > 0$, f is called ***convex***.

Case 4 This is the general case in which f is neither linear nor concave up nor concave down. Given u_l and u_r, define the linear function $l(u)$ by

$$l(u) = \frac{f(u_l) - f(u_r)}{u_l - u_r}(u - u_l) + f(u_l),$$

that is, the graph of l is the straight line through the pair of points $(u_r, f(u_r))$ and $(u_l, f(u_l))$. If the graph of f is concave up on an interval $[a, b]$ containing both u_r and u_l, then by elementary calculus,

$$l(u) > f(u)$$

whereas

$$l(u) < f(u)$$

if f is concave down.

For general f, one has the following theorem:[12] *Solutions exist, are unique, and depend continuously on the initial data (in a certain function space) if one allows only discontinuities that satisfy:*

(a) *if* $u_r > u_l$, *then* $l(u) \leq f(u)$ *for all* $u \in [u_l, u_r]$; *and*

(b) *if* $u_r < u_l$, *then* $l(u) \geq f(u)$ *for all* $u \in [u_r, u_l]$.

This is known as ***Oleinik's condition*** (E).

In summary, if f is convex (concave up or concave down), elementary calculus shows that a discontinuity allowed by the condition separates the characteristics; that is, two characteristics cross the graph of the discontinuity at each point of the (x, t) plane, and either they both point forward in time or they can both be traced backward in time. The entropy condition rules out the former possibility. One can readily verify that Oleinik's condition (E) rules out the same shocks as the entropy condition; for example, if $f_{uu} > 0$, condition (E) requires that $u_r < u_l$ across a shock. If f is not convex, condition (E) is a generalization of the entropy condition; indeed, if f is not convex, it is not obvious exactly what our earlier entropy condition does or does not allow. Condition (E) forbids in particular shocks that move faster or slower than the characteristics on both of their sides.

Notice that a discontinuity can satisfy part (a) of Oleinik's theorem without f being concave down on the whole interval $[u_l, u_r]$, as in Figure 3.4.5.

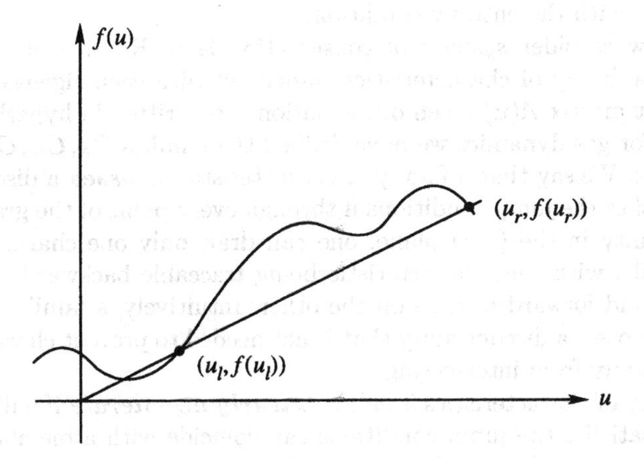

FIGURE 3.4.5. A situation obeying (a) of Oleinik's condition.

Consider next a situation in which the line joining the point $(u_l, f(u_l))$ to $(u_r, f(u_r))$ is neither wholly above nor wholly below the graph of f, so

[12] O. A. Oleinik, "Existence of Solutions for a Simple Hyperbolic Equation," *Amer. Math. Soc. Transl. Ser.* 2, **285** [1965], 33.

Oleinik's condition (E) is violated; see Figure 3.4.6.

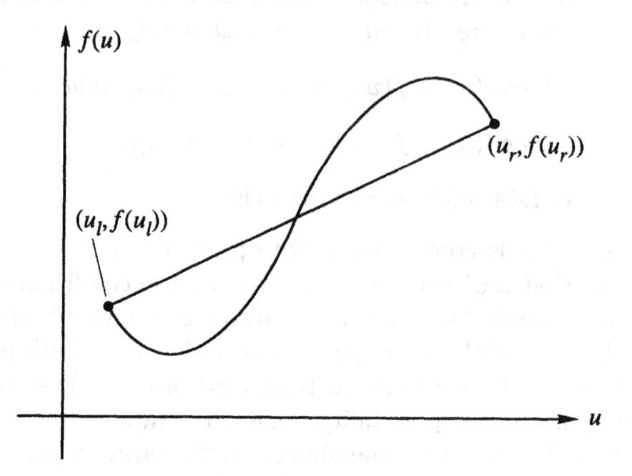

FIGURE 3.4.6. A situation violating Oleinik's condition.

Thus, we cannot connect u_r and u_l by means of a single shock. One can also show[13] that they cannot be connected by a single rarefaction fan. One can prove, however, that the gap between u_r and u_l can be bridged by a compound wave consisting of shocks, rarefaction fans, and slip lines, all consistent with the entropy condition.

We now consider *systems* of conservation laws. Recall from §**3.1** that we have a family of characteristics associated with each eigenvalue of the coefficient matrix $A(\mathbf{u})$ when our equations are written in hyperbolic form (3.1.3). For gas dynamics we have defined the families C_+, C_-, C_0 of characteristics. We say that a family of characteristics ***crosses*** a discontinuity that satisfies the jump conditions if through every point of the graph of the discontinuity in the (x,t) plane, one can draw only one characteristic of that family, with that characteristic being traceable backward in time on one side and forward in time on the other. Intuitively, a family of characteristics crosses a discontinuity that is not needed to prevent characteristics of that family from intersecting.

A family of characteristics is called ***linearly degenerate*** if a discontinuity that satisfies the jump conditions can coincide with a member of that family. For example, the C_0 family is linearly degenerate, because a slip line is also a C_0 characteristic. We said earlier that a discontinuity ***separated*** a family of characteristics if through each point of the graph of the discontinuity in the (x,t) plane there exist two characteristics, with the two of them either traceable back in time or forward in time. (Slip lines do not separate the C_0 characteristics, nor are they crossed by them.) A family

[13]See Oleinik, *Loc. cit.*

of characteristics is called ***convex*** if, whenever a discontinuity is allowed by the algebraic jump conditions, it is either crossed by the family or it separates the family. If a family of characteristics is convex, the entropy condition rules out discontinuities crossed by characteristics that can be traced forward in time. The Prandtl relation

$$u_0 u_1 = c_*^2$$

derived in §**3.2** shows that the C_+ and C_- characteristics of gas dynamics for a γ-law gas are convex families. Thus, these equations admit only two kinds of discontinuities: discontinuities that separate either the C_+ or the C_- family of characteristics, and linearly degenerate slip lines.

A system of conservation laws is said to be ***convex*** if all the discontinuities allowed by the jump conditions separate one (and only one) of the families of characteristics and if all the families are convex. A single equation can be readily seen to be convex in this sense if and only if $f(u)$ is convex. The system of isentropic gas dynamic equations (i.e., $p\rho^{-\gamma} =$ constant) is thus convex, because it has only the C_+ and C_- families of characteristics, both of which are convex.

A system of conservation laws is said to be ***linearly degenerate*** if at least one of its families of characteristics is linearly degenerate and the others are convex. Thus, *the* 3×3 *system of gas dynamics is linearly degenerate because the* C_0 *family is linearly degenerate.*

In other cases, the system is called ***nonconvex***, and compound waves are needed to connect states and construct a solution of the Riemann problem (and thus of the initial value problem). Examples of nonconvex systems are provided by the equations of gas dynamics with combustion, to which we turn next.[14]

The equations of gas dynamics with combustion are

$$\begin{aligned} \rho_t + (\rho u)_x &= 0, \\ (\rho u)_t + (\rho u^2 + p)_x &= 0, \\ e_t + ((e+p)u)_x &= 0, \end{aligned} \tag{3.4.3}$$

where

$$e = \tfrac{1}{2}\rho u^2 + \rho\epsilon \quad \text{and} \quad \epsilon = \frac{p\tau}{\gamma - 1} + q.$$

The only modification from the gas dynamic equations for a γ-law gas that we have already studied is the presence of the term q in the energy density; we have yet to specify q. The term q represents ***chemical energy*** that can change due to burning. We assume here for simplicity that q takes only two

[14]Another interesting nonconvex physical system is analyzed in A. J. Chorin, *Comm. Math. Phys.* **91** [1983], 103.

values:

$$q = \begin{cases} q_0, & \text{if the gas is unburned,} \\ q_1, & \text{if the gas is burned,} \end{cases} \tag{3.4.4}$$

where q_0 and q_1 are constants. To complete our model, we assume that the burning reaction occurs when some threshold is reached as follows: Let $T = p/\rho$ measure the ***temperature***. We assume that the gas burns when $T \geq T_c$ for a constant T_c called the ***ignition temperature***. Let $\Delta = q_1 - q_0$; the reaction is called ***exothermic*** if $\Delta < 0$, that is, $q_1 < q_0$, and is ***endothermic*** if $\Delta > 0$, that is, $q_1 > q_0$. The third equation in (3.4.3) asserts that the "total" energy is conserved, where the "total" energy is the sum of the kinetic energy, the internal energy, and a chemical energy, part of which can be released through a process we shall call "burning" or "combustion."

In this model we have neglected viscous effects, heat conduction, and radiative heat transfer. Of these the most serious is the exclusion of the effects of heat conduction.[15] We also assume that the "combustion" only occurs once; after the gas is burned, it stays burned. We have already assumed in writing (3.4.3) that we are dealing with a γ-law gas where γ is the same for both burned and unburned gas. This is not normally realistic, but the mathematics in the more general situation is essentially the same as in our simplified situation. We shall furthermore assume that the region in which combustion occurs (and in which q changes from q_0 to q_1) is infinitely thin, and that the transition occurs instantaneously.

We can picture the gas as spread along the x-axis; at any instant some sections of the gas are burned ($q = q_1$) and others are not ($q = q_0$). The lines between them will be associated with discontinuities Σ in the flow. A discontinuity across which $\Delta \neq 0$ is called a ***reaction front*** or a ***combustion front***.

The jump relations across a discontinuity Σ are given by the relations described previously. In a frame moving with the discontinuity, they read:

$$\begin{aligned} [\rho u] &= 0, \\ [\rho u^2 + p] &= 0, \\ [(e + p)u] &= 0. \end{aligned}$$

Letting $M = \rho u$, we derive

$$M^2 = \frac{p_1 - p_1}{\tau_0 - \tau_1} \tag{3.4.5}$$

[15]For the equations for flow with finite conduction and reaction rates, see F. A. Williams [1965] *Combustion Theory*, Addison-Wesley.

as before (see equation (3.2.14)). We can eliminate u_1 and u_2 from the equation to obtain a Hugoniot relation by the methods of §**3.2**. We still define the ***Hugoniot function*** with center (τ_0, p_0) to be

$$H(\tau, p) = \epsilon - \epsilon_0 + \frac{\tau_0 - \tau}{2}(p_0 + p). \tag{3.4.6}$$

so the energy jump condition becomes

$$H(\tau_1, p_l) = 0.$$

We can rewrite (3.4.6) as

$$2\mu^2 H(\tau, p) = p(\tau - \mu^2\tau_0) - p_0(\tau_0 - \mu^2\tau) + 2\mu^2\Delta,$$

where

$$\mu^2 = \frac{\gamma - 1}{\gamma + 1}.$$

This differs from our expression in §**3.2** by the term $2\mu^2\Delta$. In particular, note that a state cannot be connected to itself by a reaction front because $H(\tau_0, p_0) = \Delta$. There can be discontinuities that are not reaction fronts $(\Delta = 0)$, in which case our analysis in §**3.2** applies. For a given $\Delta \neq 0$, we still call the curve $H(\tau, p) = 0$ the ***Hugoniot curve***. It is the locus of all possible states that can be connected to the given state (τ_0, p_0).

We shall now discuss reaction fronts for exothermic reactions; that is, $\Delta < 0$. In this case the state (τ_0, p_0) lies below the hyperbola representing the Hugoniot curve, as Figure 3.4.7 shows.

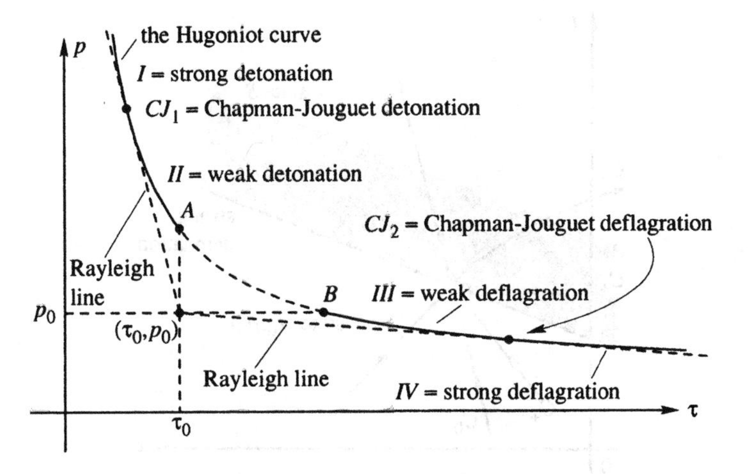

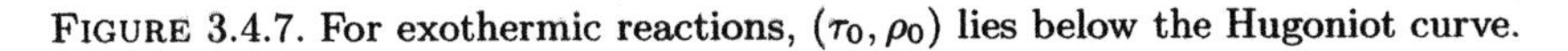
FIGURE 3.4.7. For exothermic reactions, (τ_0, ρ_0) lies below the Hugoniot curve.

One portion of the Hugoniot curve (from A to B) is omitted because the states in it correspond to negative M^2 (by (3.4.5)), which is an impossibility. The remaining part of the curve is divided into two branches. The upper branch, which corresponds to a compressive reaction front ($p > p_0$), is called the ***detonation branch***. The lower branch, which corresponds to an expansive reaction front ($p < p_0$), is called the ***deflagration branch***. The lines through (τ_0, p_0) tangent to the Hugoniot curve are called the ***Rayleigh lines***, and the points of tangency CJ_1 and CJ_2 are called the ***Chapman–Jouguet points***. These points divide the Hugoniot curve further into four subbranches. The part I above CJ_1 is called the ***strong detonation*** branch, the part II between CJ_1 and A the ***weak detonation*** branch. Similarly, the part III between B and CJ_2 is called the ***weak deflagration*** branch, the part IV below CJ_2 the ***strong deflagration*** branch. We will discuss each branch separately, and thus exclude some branches on the basis of an analogue of Oleinik's condition (E), that is, by a geometrical entropy condition.

Consider a constant state S corresponding to a point (τ_1, p_1) that lies on the Hugoniot curve, that is, it is possible to connect S and (τ_0, p_0) by a chemical reaction front. Depending on the position of S, we consider the following cases:

Case 1 S lies in the strong detonation branch, that is,

$$p_1 > p_{CJ_1}$$

(See Figure 3.4.8).

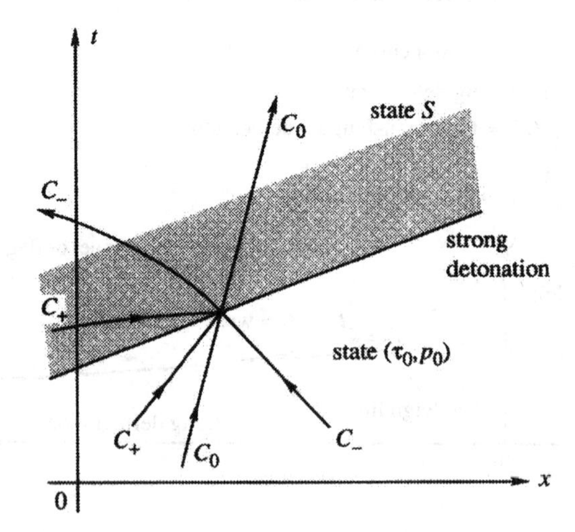

FIGURE 3.4.8. A state S on the strong detonation branch.

We can connect S to (τ_0, p_0) by a strong detonation front. One can show by the argument in §**3.2** that the gas flow relative to the reaction front is supersonic in the front and subsonic in the back, that is,

$$|u_0| > |u_1| < c_1.$$

Therefore, if the strong detonation front faces in the positive x direction, it separates the C_+ characteristics, as the figure shows. The other two families of characteristics cross the reaction front. Thus, the geometry of the characteristics in the strong detonation case is similar to what it is in the case of a shock. Hence, the pressure in the back state is sufficient to determine the back state from the front state.

Case 2 $S = CJ_1$, that is,

$$p_1 = p_{CJ_1}$$

(See Figure 3.4.9).

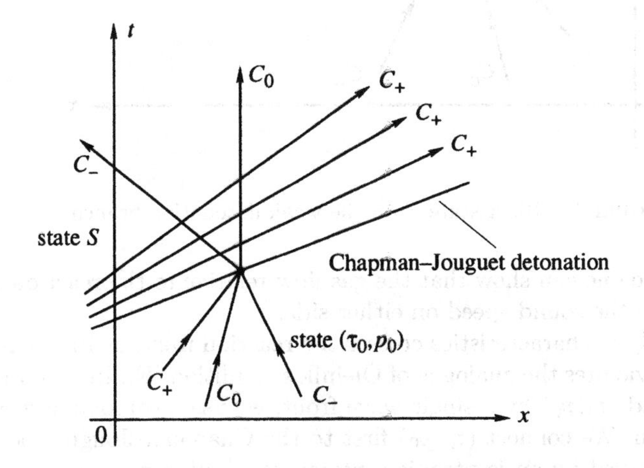

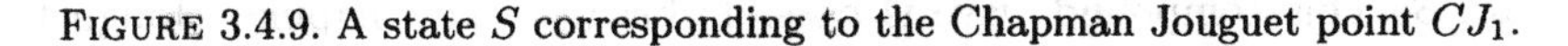

FIGURE 3.4.9. A state S corresponding to the Chapman Jouguet point CJ_1.

In this case the velocity relative to the reaction front is supersonic in the front and sonic in the back, that is,

$$|u_0| > c_0 \quad \text{and} \quad |u_1| = c_1.$$

Thus, the Chapman–Jouguet reaction front, when observed from the burned gas state, is one of the characteristics (of C_+ or C_- family). The condition $|u_1| = c_1$ enables us to determine the back state from the front state without further assumption.

Case 3 S lies in the weak detonation branch, that is,

$$p_{CJ_1} > p_1 \geq p_A$$

(See Figure 3.4.10).

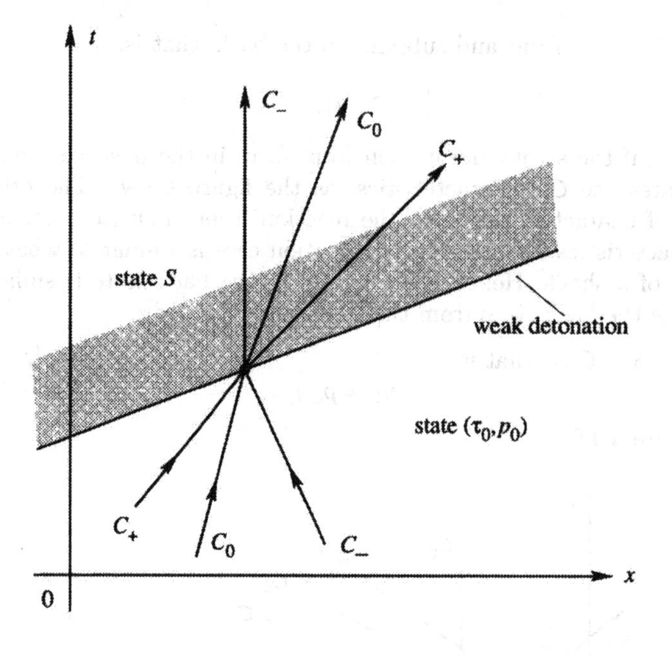

FIGURE 3.4.10. A state S on the weak detonation branch.

In this case one can show that the gas flow relative to the reaction front is faster than the sound speed on either side.

Each family of characteristics crosses the reaction front, as Figure 3.4.10 shows. This violates the analogue of Oleinik's condition (E). Instead of connecting S and (τ_0, p_0) by a single wave front, we use a compound wave to connect them. We connect (τ_0, p_0) first to the Chapman–Jouguet detonation and followed by an isentropic-centered rarefaction wave to reach the state S. This compound wave is possible because the Chapman–Jouguet detonation moves with sound speed with respect to the gas flow in its back side. If we adopt this compound wave to connect them, the pressure is S and the state (τ_0, p_0, u_0) will determine S completely. If weak detonations are not excluded, the solution of the initial value problem is not unique; indeed, the state S can then be connected to a state with a pressure p_0 by either a weak detonation or a CJ detonation followed by a rarefaction. The exclusion of the weak detonation is analogous to the exclusion of discontinuities that move faster than the characteristics on both their sides by means of condition (E).

Case 4 S lies in the weak deflagration branch, that is,

$$p_0 \geq p_1 > p_{CJ_2}$$

(See Figure 3.4.11).

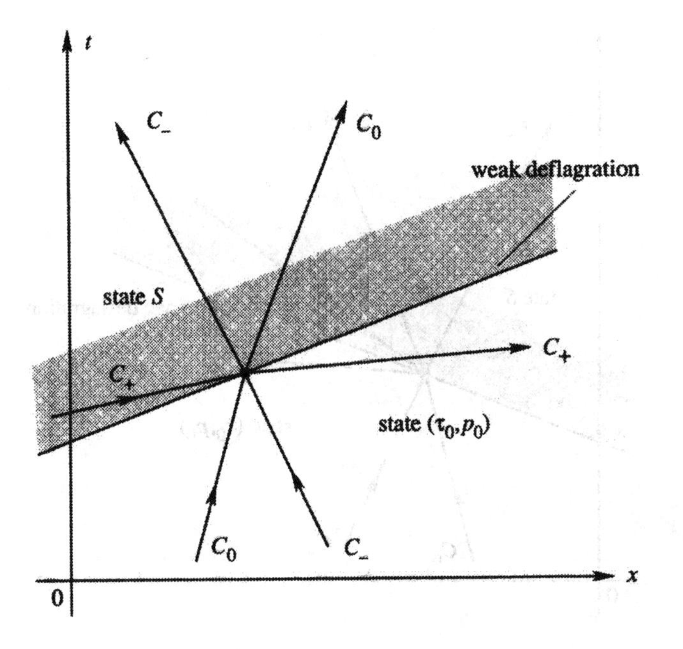

FIGURE 3.4.11. A state S on the weak deflagration branch.

One can show that the reaction front moves with respect to the gas slower than the sound speed on both sides. This is an indeterminate case. One can determine the solution uniquely only by taking into account heat conduction and a finite reaction rate. In the limiting case of no heat conduction (which we have assumed here), it can be shown[16] that the only weak deflagration that can occur is one across which the pressure is constant and there is no mass flow, that is, this deflagration is indistinguishable from a slip line. Thus, only in the case where $p_1 = p_0$ can the state S be connected to (τ_0, p_0) by a weak deflagration.

Case 5 S lies in the strong deflagration branch, that is,

$$p_1 < p_{CJ_2}$$

(See Figure 3.4.12).

Then the gas flow relative to the reaction front is subsonic in the front side and supersonic in the back side. If this strong deflagration moves in the positive x direction, it separates the C_+ characteristics. However, this separation does not satisfy the geometric entropy condition, although the strong deflagration is consistent with the conservation laws. Thus, we must exclude strong deflagrations.

---

[16]See A. J. Chorin, "Random choice methods with applications to reacting gas flow," *J. Comp. Phys.*, **25** [1977], 253.

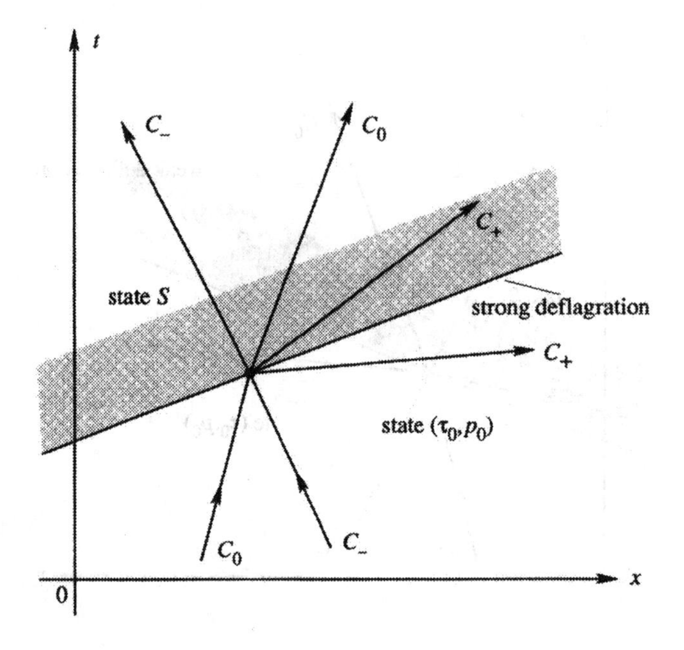

FIGURE 3.4.12. A state S on the strong deflagration branch.

Note that we have excluded weak detonations and strong detonations by means of plausible geometrical entropy conditions; neither one could be excluded by the requirement that the physical entropy should increase as the gas crosses the discontinuity. They could, of course, be excluded by other considerations of a plausible physical nature. (The theory of such waves is still in a state of development.)[17]

These considerations permit us to propose a solution for the Riemann problem for the system (3.4.3). We define a ***centered wave*** to be either a shock, a centered rarefaction, a strong detonation, or the compound wave consisting of the Chapman–Jouguet detonation followed by a centered rarefaction. A ***Riemann problem*** for the system (3.4.3) is an initial value problem for (3.4.3) with initial data of the form

$$\begin{bmatrix} \rho \\ u \\ e \\ q \end{bmatrix} = S_r, \quad \text{if } x \geq 0;$$

[17]A. Bourlioux, A. Majda, and V. Roytburd, "Theoretical and numerical structures for unstable one dimensional detonations," *SIAM J. Appl. Math.*, **51**,[1991], 303–343. P. Colella, A. Majda, and V. Roytburd, "Theoretical and numerical structure of reactive shock waves," *SIAM J. Sci. Comput.*, **1**, [1980], 1059–1080.

and

$$\begin{bmatrix} \rho \\ u \\ e \\ q \end{bmatrix} = S_l, \quad \text{if } x < 0,$$

where S_r and S_l are two constant states of the gas. We claim that the Riemann problem is solvable by connecting S_r to S_l through a right-centered wave, a constant state I, a slip line, a constant state II, and a left-centered wave (Figure 3.4.13)

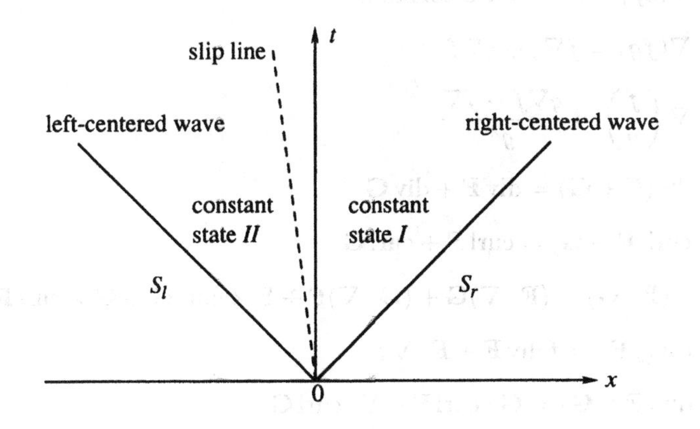

FIGURE 3.4.13. The Riemann problem for the gas flow with combustion.

The argument is similar to the case of an inert gas (§**3.3**). If we can determine the constant state I from S_r and a given pressure p_* in I, we can proceed as in §**3.3** to get an algebraic equation for p_* through the continuity of the velocity across the slip line, and hence obtain a solution to the Riemann problem. For this purpose, we assume first that S_r is in an unburned state. If constant state I is in an unburned state, we are back to the case of inert gas. If constant state I is in a burned state, we can connect I to S_r by a centered wave depending on the position of p_* in the Hugoniot curve with center S_r. The critical criterion is whether the temperature T computed in the constant state I exceeds the ignition temperature T_c or not. One can actually show that there is a consistent way of solving the Riemann problem.[18] If S_r contains burned gas, the constant state I also contains burned gas, because the burning can occur only once.

Having solved the Riemann problem, we can use the random choice method described in §**3.2** and §**3.3** to obtain the approximate solutions

[18]See A. J. Chorin, *Random choice methods with applications to reacting gas flow*, J. Comp. Phys., **25** [1977], 253 for the details.

for the general initial value problem for the system (3.4.3). Results similar to those in §**3.3** can be expected. We need not repeat the construction.

Vector Identities

The following gives some general formulas that are useful for calculations with vector fields in $\mathbb{R}^3$.

1. $\nabla(f+g) = \nabla f + \nabla g$
2. $\nabla(cf) = c\nabla f$, for a constant c
3. $\nabla(fg) = f\nabla g + g\nabla f$
4. $\nabla\left(\dfrac{f}{g}\right) = \dfrac{g\nabla f - f\nabla}{g^2}$
5. $\mathrm{div}(\mathbf{F}+\mathbf{G}) = \mathrm{div}\,\mathbf{F} + \mathrm{div}\,\mathbf{G}$
6. $\mathrm{curl}(\mathbf{F}+\mathbf{G}) = \mathrm{curl}\,\mathbf{F} + \mathrm{curl}\,\mathbf{G}$
7. $\nabla(\mathbf{F}\cdot\mathbf{G}) = (\mathbf{F}\cdot\nabla)\mathbf{G} + (\mathbf{G}\cdot\nabla)\mathbf{F} + \mathbf{F}\times\mathrm{curl}\,\mathbf{G} + \mathbf{G}\times\mathrm{curl}\,\mathbf{F}$
8. $\mathrm{div}(f\mathbf{F}) = f\,\mathrm{div}\,\mathbf{F} + \mathbf{F}\cdot\nabla f$
9. $\mathrm{div}(\mathbf{F}\times\mathbf{G}) = \mathbf{G}\cdot\mathrm{curl}\,\mathbf{F} - \mathbf{F}\cdot\mathrm{curl}\,\mathbf{G}$
10. $\mathrm{div}\,\mathrm{curl}\,\mathbf{F} = 0$
11. $\mathrm{curl}(f\mathbf{F}) = f\,\mathrm{curl}\,\mathbf{F} + \nabla f\times\mathbf{F}$
12. $\mathrm{curl}(\mathbf{F}\times\mathbf{G}) = \mathbf{F}\ \mathrm{div}\ \mathbf{G} - \mathbf{G}\ \mathrm{div}\ \mathbf{F} + (\mathbf{G}\cdot\nabla)\mathbf{F} - (\mathbf{F}\cdot\nabla)\mathbf{G}$
13. $\mathrm{curl}\ \mathrm{curl}\,\mathbf{F} = \mathrm{grad}\,\mathrm{div}\ \mathbf{F} - \nabla^2\mathbf{F}$
14. $\mathrm{curl}\,\nabla f = 0$
15. $\nabla(\mathbf{F}\cdot\mathbf{F}) = 2(\mathbf{F}\cdot\nabla)\mathbf{F} + 2\mathbf{F}\times(\mathrm{curl}\,\mathbf{F})$
16. $\nabla^2(fg) = f\nabla^2 g + g\nabla^2 f + 2(\nabla f\cdot\nabla g)$
17. $\mathrm{div}(\nabla f\times\nabla g) = 0$
18. $\nabla\cdot(f\nabla g - g\nabla f) = f\nabla^2 g - g\nabla^2 f$
19. $\mathbf{H}\cdot(\mathbf{F}\times\mathbf{G}) = \mathbf{G}\cdot(\mathbf{H}\times\mathbf{F}) = \mathbf{F}\cdot(\mathbf{G}\times\mathbf{H})$
20. $\mathbf{H}\cdot((\mathbf{F}\times\nabla)\times\mathbf{G}) = ((\mathbf{H}\cdot\nabla)\mathbf{G})\cdot\mathbf{F} - (\mathbf{H}\cdot\mathbf{F})(\nabla\cdot\mathbf{G})$
21. $\mathbf{F}\times(\mathbf{G}\times\mathbf{H}) = (\mathbf{F}\cdot\mathbf{H})\mathbf{G} - \mathbf{H}(\mathbf{F}\cdot\mathbf{G})$

Notes.

In identity 7, $\mathbf{V} = (\mathbf{F} \cdot \nabla)\mathbf{G}$ has components $\mathbf{V}_i = \mathbf{F} \cdot (\nabla G_i)$, for $i = 1, 2, 3$, where $\mathbf{G} = (G_1, G_2, G_3)$.

In identity 13, the vector field $\nabla^2 \mathbf{F}$ has components $\nabla^2 F_i$, where $\mathbf{F} = (F_1, F_2, F_3)$.

In identity 20, $(\mathbf{F} \times \nabla) \times \mathbf{G}$ means ∇ is to operate only on $\mathbf{G}$ in the following way: To calculate $(\mathbf{F} \times \nabla) \times \mathbf{G}$, we define $(\mathbf{F} \times \nabla) \times \mathbf{G} = \mathbf{U} \times \mathbf{G}$ where we define $\mathbf{U} = \mathbf{F} \times \nabla$ by:

$$\mathbf{U} = \mathbf{F} \times \nabla = \begin{vmatrix} \mathbf{i} & \mathbf{j} & \mathbf{k} \\ F_1 & F_2 & F_3 \\ \dfrac{\partial}{\partial x} & \dfrac{\partial}{\partial y} & \dfrac{\partial}{\partial z} \end{vmatrix}.$$

Index

A
acceleration of a fluid particle, 4
algorithm, 94
almost potential flow, 59
asymptotically stable, 97
autonomous, 97

B
back, 124
balance of momentum, 4, 6, 12
  differential form, 6
  integral form, 7
Bernoulli's theorem, 16, 48, 55, 71
bifurcation, 99, 100
Blasius' theorem, 52
body
  force, 6
  force on, 52
boundary
  condition, 34, 41
  layer, 67, 68
    approximation, 76
    equation, 75
    separation, 79
    thickness, 71
    vorticity in, 76
  layer separation, 71
burning, 152

C
Cauchy's theorem, 32
centered
  rarefaction wave, 114
  wave, 137, 138, 158
central limit theorem, 84
channel flow, 17
Chapman–Jouguet points, 154
characteristic, 103, 105, 108
  intersecting, 106, 108
  length, 35
  linearly degenerate, 150
  velocity, 35
chemical energy, 151
circulation, 21, 48, 57
coefficients of viscosity, 33
combustion, 103, 152
  front, 152
  wave, 145
complex
  potential, 51

variables methods, 51
velocity, 51, 53
compressible
flow, 14, 33, 103
compressive shock, 131, 134
concave up, 146
conformal transformation, 65
connected
state, 115
wave, 115
conservation
law, 122, 129, 145
of energy, 12
of mass, 2, 12
of vorticity, 28
consistency of an algorithm, 94
constant state, 111
contact discontinuity, 124
continuity equation, 3
continuum assumption, 2
convective term, 40
convergence of an algorithm, 94
convex characterisitics, 151
Couette flow, 31
Courant–Friedrichs–Lewy condition, 141
cylindrical coordinates, 45

D

d'Alembert's paradox, 54
in 3d, 58
decompostion theorem, 37
deflagration branch, 154
deformation, 18
tensor, 19, 31
degenerate
linearly, 151
density, 1, 14
derivative
material, 5
detonation branch, 154
differential
form, 120
form of mass conservation, 3
diffusion, 39
discontinuity, 121, 146, 148
separating, 148
disjoint, 82
event, 82
dissipation term, 39
divergence
free part, 38
space-time, 119
downstream, 93
drag, 54, 57, 60, 66
form, 80
skin, 80
dynamics, 96

E

endothermic, 152
energy, 12
equation, 118
flux, 18
internal, 12, 117
kinetic, 12
per unit volume, 117
enthalpy, 14
entropy, 14, 118, 158
condition, 127, 130, 133, 157
equation
continuity, 3
differential form, 120
Euler, 13, 15
heat, 84
Hugoniot, 125
Navier–Stokes, 34
of state, 44
Prandtl, 75
Stokes', 40
vorticity, 24
weak form, 120
error function, 70
Euler equation, 13, 15, 49, 78, 94, 96
event, 82
disjoint, 82
exothermic, 152
expectation, 82, 143

F

field
- velocity, 1
- vorticity, 18

filament, 65
Filon's paradox, 67
first
- coefficient of viscosity, 33
- law of thermodynamics, 14, 118

fixed point, 97
flat plate, 69
flow, 14
- almost potential, 59
- around a disk, 55
- around a half circle, 51
- around an obstacle, 52
- between plates, 41
- between two plates, 42
- compressible, 14, 33
- Couette, 31
- gas, 103
- homogeneous, 11, 48
- ideal, 48
- in a channel, 17
- in a pipe, 45
- in the half-plane, 69
- in the upper half-plane, 51
- incompressible, 10
- induced by a vortex filament, 65
- irrotational, 47
- isentropic, 14, 15
- map, 7, 95
- over a plate, 66
- past a sphere, 59
- Poiseuille, 45
- potential, 47, 48
- potential around a disk, 55
- potential vortex, 56
- stationary, 29, 49
- supplementary region half-plane, 51

fluid
- flow map, 7
- ideal, 5
- particle, 4
- velocity, 1
- viscous, 33

flux, 7
- of vorticity, 22

force, 5, 53
- on a body, 52

form drag, 80
fourth power law, 46
front, 124
function
- error, 70
- Green's, 61, 86

G

gamma
- law gas, 114, 118, 125, 131, 139
- simple wave, 111

gas
- dynamics, 103, 111
- flow, 103
- ideal, 118, 125, 131, 139

Gaussian, 84
generation of vorticity, 43
geometric entropy condition, 135
Glimm's existence proof, 145
global stability, 97
gradient part, 38
Green's function, 61, 63, 64, 86

H

half-plane flow, 69
Hamiltonian system, 62
heat equation, 84, 86
Helmholtz
- decomposition theorem, 37
- theorem, 26, 37

Hodge theorem, 37
hodograph transformation, 110
homogeneous, 11
- flow, 48

Hopf bifurcation theorem, 99
Hugoniot

curve, 126, 153
equation, 125, 139
function, 125, 153
hyperbolic, 104

I
ideal
flow, 48
fluid, 5, 31
gas, 44, 118, 125, 131, 139
ignition temperature, 152
incompressible, 11
approximately, 44
flow, 10, 13, 34
independent, 82
random variables, 83
integral
form, 120
form of balance of momentum, 7
form of mass conservation, 3
intensity of a vortex sheet, 88
internal energy, 12, 40, 117
invariant
Riemann, 109
irrotational, 47
isentropic
flow, 14, 15
fluids, 14
gas flow, 122

J
Jacobian, 8
matrix, 24
jump, 122
discontinuity, 121, 122
relations, 124

K
Kelvin's circulation theorem, 21
kinematic viscosity, 34
kinetic energy, 12, 40, 49
Kutta–Joukowski theorem, 53

L
law
conservation, 129, 145
of large numbers, 83
layer
boundary, 67, 71
leading edge, 93
left
-centered wave, 138
state, 115
length
characteristic, 35
Liapunov stability theorem, 97
Lie derivative, 43
Lie–Trotter product formula, 95
line
vortex, 22
linearly degenerate, 150, 151

M
Mach number, 44
mass
conservation, 11
density, 1
flow rate, 46
matching solutions, 78
material derivative, 5, 18
mean, 82
mechanical jump relations, 124
momentum
balance of, 6
flux, 7
moving with the fluid, 7

N
Navier–Stokes equation, 31, 33, 38, 67, 77, 94, 95
Neumann problem, 37, 49, 63
Newton's second law, 2, 6
no-slip condition, 34, 43
noncompressive shock, 133
nonconvex, 151
nonlinear dynamics, 96

O
obstacle, 58
flow around, 52

Oleinik's condition, 149, 154
orthogonal projection, 38
oscillations, 100
Oseen's equation, 66

P
paradox
  d'Alembert, 54, 58
  Filon, 67
  Stokes, 67
pipe flow, 45
piston, 111
plate, 80
  flow between, 41
  flow over, 66, 69
  flow past, 87
point vortices, 60
Poiseuille flow, 45
potential
  complex, 51
  flow, 47, 48, 51, 56, 58
    almost, 59
  flow around a disk, 55
  velocity, 48
  vortex, 56
  vortex flow, 56
Prandtl
  boundary layer equation, 73, 75
  equation, 78, 94
  relation, 132, 151
pressure, 5, 14
probability, 82
  density function, 83
  theory, 82
product formula, 95
projection operator, 38

Q
quasilinear, 104

R
random
  choice method, 144, 159
  variable, 82, 141
  Gaussian, 84
  walk, 85, 88, 96
Rankine–Hugoniot relations, 124
rarefaction
  fan, 148
  shock, 127, 129
  wave, 114
Rayleigh lines, 154
reaction
  endothermic, 152
  exothermic, 152
  front, 152
reversibility, 136
Reynolds number, 36, 96
Riemann
  invariant, 109, 110, 113
  problem, 103, 137, 139, 144, 158, 159
right
  -centered wave, 138
  state, 115
rigid rotation, 18
rotation, 18

S
sample space, 82
scaling argument, 81
second
  coefficient of viscosity, 33
  law, 118
separate characteristics, 130
separated, 150
separation
  boundary, 68, 71, 79
shadow of a vortex sheet, 90
sheet
  vortex, 22
shock, 117, 124
  back, 124
  compressive, 131, 133
  front, 124
  noncompressive, 133
  rarefaction, 127, 129
  separating, 130
  tube problem, 137

similar flows, 36
simple wave, 111
simply connected, 47
skin drag, 80
slightly viscous flow, 47
slip line, 124, 138, 140
sound speed, 44, 104, 131
space-time divergence, 119
spatial velocity field, 1
speed
    discontinuity, 147
sphere
    flow past, 59, 66
stability, 96
stable point, 97
stagnation point, 29, 55
standard deviation, 83
state, 111, 136, 137
    connected, 115
    equation of, 44
stationary, 49
    flow, 16, 58
    flow criterion, 29
steady flow, 58
Stokes
    equation, 40, 67, 96
    flow, 66
    paradox, 67
stream function, 29, 43, 54, 80
streamline, 16, 29, 44
strength of a vortex tube, 26
stress tensor, 32
stretched, 27
strong
    deflagration, 154
    detonation, 154
subsonic, 131, 157
supersonic, 131
symmetry, 101

T
tangential
    boundary condition, 34
    forces, 5
Tchebysheff's inequality, 84
temperature, 14, 152
test functions, 119
theorem
    Bernoulli's, 16
    Blasius', 52
    Cauchy's, 32
    central limit, 84
    Helmholtz, 26
    Helmholtz–Hodge, 37
    Kelvin circulation, 21
    Kutta–Joukowski, 53
    transport, 10
thermodynamics, 14, 118
    first law, 14, 118
thickness
    boundary layer, 71, 73
total force, 8
trajectory, 16
transfer of momentum, 31
transformation
    hodograph, 110
translation, 18
transport theorem, 10, 18, 117
tube
    vortex, 26

U
upper half-plane
    flow in, 51

V
variance, 83, 143
variation, 129
velocity
    characteristic, 35
    complex, 51
    field, 1
    potential, 48, 51
    profile, 42
viscosity, 129
    coefficients, 33
    kinematic, 34
viscous fluid, 33
vortex, 61
    filament, 65

line, 22
potential, 56
sheet, 22, 82, 87
sheet intensity, 88
tube, 26
tube, strength of, 26
vortices
point, 60
vorticity, 18, 63
conservation, 28
creation operator, 96
equation, 28
generation of, 43
in boundary layer, 76
transport, 23

W
wave
centered, 137
connected, 115
left-centered, 138
rarefaction, 114
right-centered, 138
simple, 111
weak
deflagration, 154
detonation, 154
solution, 118, 119

Texts in Applied Mathematics

(continued from page ii)

33. *Thomas:* Numerical Partial Differential Equations: Conservation Laws and Elliptic Equations.
34. *Chicone:* Ordinary Differential Equations with Applications.
35. *Kevorkian:* Partial Differential Equations: Analytical Solution Techniques, 2nd ed.
36. *Dullerud/Paganini:* A Course in Robust Control Theory: A Convex Approach.
37. *Quarteroni/Sacco/Saleri:* Numerical Mathematics.

9781461269342.